SpringerBriefs in Materials

The SpringerBriefs Series in Materials presents highly relevant, concise monographs on a wide range of topics covering fundamental advances and new applications in the field. Areas of interest include topical information on innovative, structural and functional materials and composites as well as fundamental principles, physical properties, materials theory and design. **Indexed in Scopus (2022).**

SpringerBriefs present succinct summaries of cutting-edge research and practical applications across a wide spectrum of fields. Featuring compact volumes of 50 to 125 pages, the series covers a range of content from professional to academic. Typical topics might include

- A timely report of state-of-the art analytical techniques
- A bridge between new research results, as published in journal articles, and a contextual literature review
- A snapshot of a hot or emerging topic
- An in-depth case study or clinical example
- A presentation of core concepts that students must understand in order to make independent contributions

Briefs are characterized by fast, global electronic dissemination, standard publishing contracts, standardized manuscript preparation and formatting guidelines, and expedited production schedules.

Mohammad M. Farag · Zainab M. Al-Rashidy

Biomaterials for Tissue Regeneration

Advances and Challenges for Fabrication
and Clinical Translation

 Springer

Mohammad M. Farag
Glass Research Department (Biomaterials
Group)
National Research Centre
Giza, Egypt

Zainab M. Al-Rashidy
Refractories, Ceramics and Building
Materials (Biomaterials Group)
National Research Centre
Giza, Egypt

ISSN 2192-1091 ISSN 2192-1105 (electronic)
SpringerBriefs in Materials
ISBN 978-3-031-75753-2 ISBN 978-3-031-75754-9 (eBook)
https://doi.org/10.1007/978-3-031-75754-9

This Springer imprint is published by the registered company Springer Nature Switzerland AG
The registered company address is: Gewerbestrasse 11, 6330 Cham, Switzerland

If disposing of this product, please recycle the paper.

To the soul of our dear mothers ... May God have mercy on them.
To the soul of our dear Fathers ... May God have mercy on them.
To Dr. Farag's wife and life partner
To Dr. Al-Rashidy's husband and life supporter
To our dear son "Ziad" زياد *and daughter "Salma"* سلمى
To all our families inside and outside our beloved homeland
To all researchers and students of knowledge We dedicate to you the fruits of my effort on my book "Biomaterials for Tissue Regeneration—Advances and Challenges for Fabrication and Clinical Translation"

Acknowledgments

At the beginning of the speech, Dr. Farag and Dr. Zainab Al-Rashidy must first thank God who enabled us to complete this book.

Dr. Farag and Dr. Zainab Al-Rashidy also extend their thanks and gratitude to everyone:

Soul of our dear fathers, the soul of our honorable mothers, Dr. Farag's dear wife, and Dr. Al-Rashidy's dear husband who were the first to support us in reaching to achieve this book.

We also extend our thanks and gratitude to our son, daughter, brothers, and sisters.

We can only extend our sincere thanks to our colleagues in our work.

Contents

Abbreviations

3D	Three-dimensional
3DP	3D printing
ACP	Amorphous calcium phosphate
AD	Latin phrase "Anno Domini", it means years after the birth of Jesus
AI	Artificial intelligence
AM	Additive manufacturing
ATZ	Alumina toughened zirconia
BC	Before civilization
BG	Bioactive glass
BMMSC	Bone marrow-derived mesenchymal stem cells
CAD	Computer-aided design
CaP	Calcium phosphate
CDHAp	Calcium deficient hydroxyapatite
CNN	Convolutional neural network
CPCs	Calcium phosphate cements
DCPA	Dicalcium phosphate anhydrous
DCPD	Dicalcium phosphate dihydrate
dECM	Decellularized extracellular matrix
DL	Deep learning
DNA	Deoxyribonucleic acid
ECM	Extracellular matrix
FDA	Food and drug administration
FDM	Fused deposition modeling
hADSCs	Human adipose-derived stem cells
HAp	Hydroxyapatite
IB	Bioactivity index
MCPA	Monocalcium phosphate anhydrous
MCPM	Monocalcium phosphate monohydrate
MgP	Magnesium phosphate
micro-CT	Micro-computed tomography
ML	Machine learning

MPCs	Magnesium phosphate cements
MRI	Magnetic resonance imaging
MSCs	Mesenchymal stem cells
nBG	Nanobioactive glass
NIH	National Institutes of Health
OCP	Octacalcium phosphate
OOC	Organ-on-a-chips
OXA	Oxyapatite
PCL	Poly(ε-caprolactone)
PDGF	Platelet-derived growth factor
PEEK	Polyether ether ketone
PEG	Polyethylene glycol
PGA	Poly(glycolic acid)
PLGA	Poly(L-lactic-co-glycolic acid)
PLLA	Poly(L-lactic acid)
PMMA	Poly(methyl methacrylate)
PU	Polyurethane
PVA	Polyvinyl alcohol
RNA	Ribonucleic acid
RP	Rapid prototyping
SBF	Simulated body fluid
SFF	Solid freeform
SLA	Stereolithography
SLS	Selective laser sintering
TE	Tissue engineering
TetCP	Tetracalcium phosphate
TG	Tragacanth
α-TCP	α-Tricalcium phosphate
β-TCP	β-Tricalcium phosphate

Chapter 1
Introduction

Abstract This chapter presents the widely accepted definition of biomaterials. Moreover, the history and chronological evolution of biomaterials in ascending history are as follows, before civilization (BC), Ancient Egyptians, Romans & Greeks, Muslims, middle centuries, and today, and showing the characteristic biomaterials and biomedical devices for each period. The impact of biomaterial tissue engineering is demonstrated concerning the progressive increase number of publications. Finally, accepted tissue engineering definitions and scaffold ideal properties are shown.

1.1 Definition of Biomaterials

There have been several definitions stated for biomaterials based on the updated development of materials science, but the definition employed by the National Institutes of Health (NIH) is comprehensive and recently accepted by material scientists. According to NIH biomaterial is defined as "any natural or synthetic substance or combination of substances, other than drugs, which can be used to augment partially or whole replace any tissue, organ or function of the body, to maintain or improve quality of life of individual".

1.2 History of Biomaterials

Everything in this world is undergoing and has undergone evolution as a result of the development of science, human knowledge and skills, and the search for a better quality of life. Biomaterials are one of these things that have gone through many stages of development during human civilization. Where synthetic materials have been used for thousands of years to replace parts of the human body. Table 1.1 shows a brief of distinctive biomaterials used in each era. In early civilization, Ancient Egyptians used linen sutures for wound healing, metal sutures for teeth loose, and

Table 1.1 Distinctive biomaterials are used in each era of humankind civilization

Era	Distinctive biomaterials
Today	Tissue regeneration using scaffolds and living cells
Middle centuries	• Metallic suture: ligatures of gold wire • Copper and bronze dental and bone implants • Tin dental fillings
Muslims	• Suture from animal gut • Bone screws • New surgical instruments
Roman & Greek	Silk and gold wires to fix teeth to the gum
Ancient Egyptians	• Linen sutures • Wooden toe
Before civilization	Nacre teeth from sea shells (Mayan people)

wood for prostheses such as wooden toe discovered in Egyptian mummy (Fig. 1.1). In progress, an iron dental implant found in a dead body is returning in 200 AD. Copper and bronze dental and bone implants were used in the pre-Christian era until the nineteenth century, and so, metallic prostheses made with alloys were used for a long time. During Islamic civilization, so many instruments have been developed and created to the degree that recent surgical tools are similar to some of Muslims' surgical tools [1]. For example, Al-Zahrawi, one of the pioneer surgeons of the tenth century in the Muslims period in Spain (Al-Andalus) and is called the father of modern surgery, created about 200 surgical instruments alone, among them bone screws, sutures (from animal gut), and forceps. Figure 1.2 shows some of the surgical instruments invented and developed by Al-Zahrawi. The contributions of Al-Zahrawi in surgery and biomedical engineering fields are as much as cannot listed here, but the reader can refer to reference [2]. From sixteen to nineteen centuries, tin dental fillings, artificial teeth (made mainly from silver, copper, ivory, porcelain, etc.), and amalgam have been used as dental restorative materials [3]. However, most biomaterials before 1950 did not apply the concept of biocompatibility and sterilization. After this time, these concepts became more understandable and taken into consideration during the design and improving new biomaterials. Besides, the discovery of polymer materials since the 1930s has effectively helped to discover and develop new biomaterials with distinctive mechanical properties. For instance, polymethyl methacrylate (PMMA) has been used in dentistry and metallic orthopedic implant adhesive.

1.3 Evolution of Biomaterials

After principally knowledge of biocompatibility, biomaterials have been developed quickly and they passed through several stages. These stages can be subdivided into three generations of biomaterials, while some scientists divide them into four

Fig. 1.1 Ancient Egyptian
wooden toe prosthesis was
designed. Photo by
University of Basel

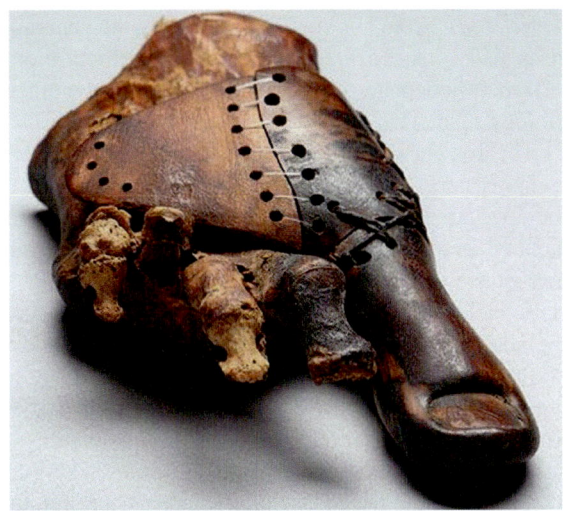

Fig. 1.2 Simulation of some
of the surgical instruments
invented by Al-Zahrawi [4]

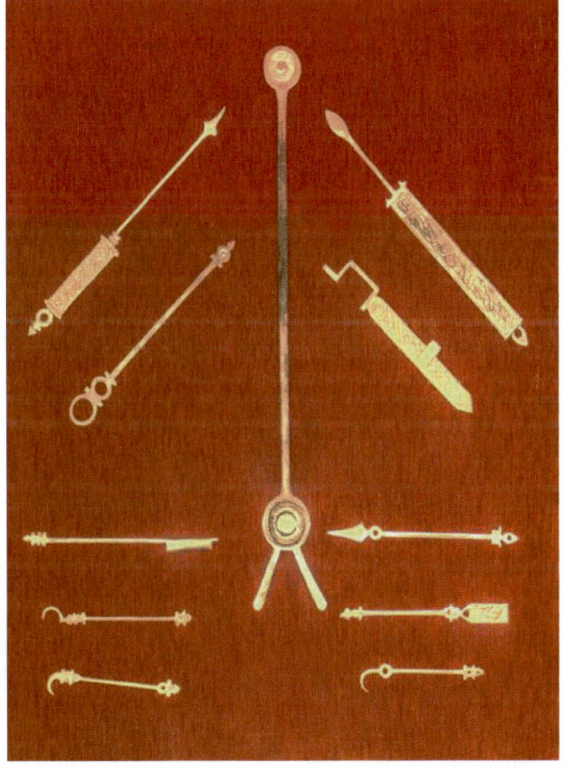

generations. From our view, we are seeing the generation adjacent to the third generation will consist of tissue regeneration without biomaterials, and its development will mainly depend on the advancement of manufacturing techniques and the cell biology field. This will be discussed in Chap. 5.

The first generation of biomaterials is bioinert materials, such as alumina, metals, and polymers (Fig. 1.3). These materials are not able to form an interface bond with the implanted tissue, but when these materials are implanted in the body, they are isolated and encapsulated with a non-adherent fibrous tissue when it implanted as a response to the body. The second generation of biomaterials is bioactive materials, such as hydroxyapatite and glasses. Herein, chemical bonds can be formed at the material-tissue interface and there is usually no immunological response of the body towards such materials. The third generation of biomaterials is biomimetic and regenerative materials which resemble the structure and function of certain tissues. This kind of material is designed to stimulate cell proliferation and differentiation using bioactive and bioresorbable substrates with interconnected porous structures to regenerate the tissue. 3D interconnected porous substrate, known as scaffold, is the success key of tissue reconstruction, and it can be combined with cells, growth factors, and biomolecules. Accordingly, the tissue engineering scaffold field has been released as a promising approach for tissue reconstruction.

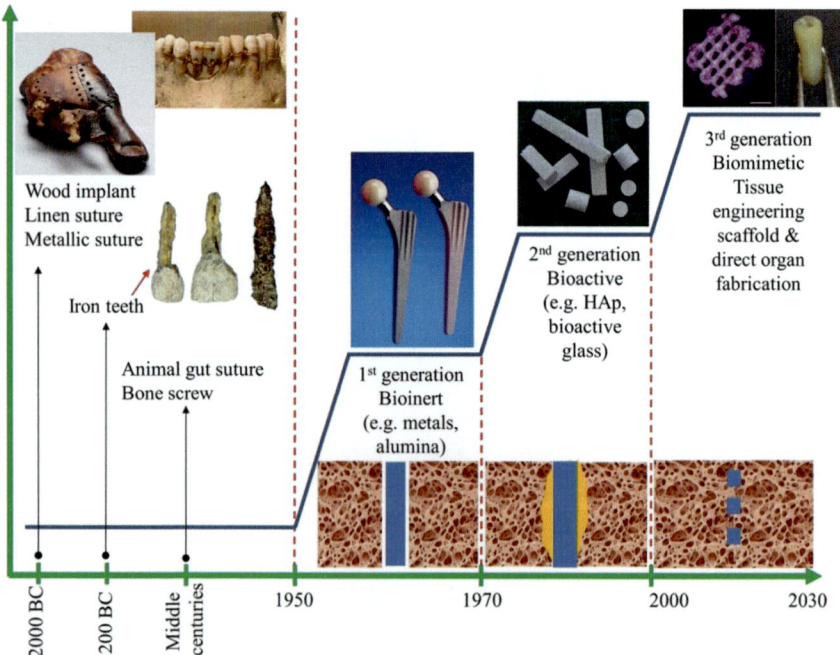

Fig. 1.3 Evolution of biomaterials started before civilization to recent. Since 1950, biomaterials have shown quick development to a degree that 3 generations released. 1st generation is bioinert material, 2nd generation is bioactive material, 3rd generation is biomimetic and regenerative material

Research and development of biomaterials used for tissue regeneration have progressively increased during the last 25 years, as shown in Fig. 1.4. Where, the total number of papers published on biomaterials for tissue regeneration topic is 34,681 from the year 2000 till now (Fig. 1.3a), moreover, the most relevant scientific field to tissue regeneration is materials science (biomaterials) followed by engineering. These facts are elicited from the Scopus database by using the search words "biomaterials" and "tissue engineering". Meanwhile, replacing the word "tissue engineering" with "tissue regeneration" ("biomaterials" word was used in both cases) gave a lower number of publications (13,607 publications). However, the biomaterials subject area possessed the highest number of publications in both cases. Despite tissue engineering and tissue regeneration, terms are used synonymously, there is a little bit of difference between them. Anywhere, tissue engineering includes scaffolds, cells, and growth factors to regenerate tissues, while tissue regeneration includes tissue engineering with specific other approaches, consisting of gene therapy, cell-based therapy, and immunomodulation, to stimulate in vivo tissue regeneration [5]. Nevertheless, the number of publications showed that the biomaterials field is an important branch of science for tissue regeneration discipline, and it is considered the key to success in developing tissue engineering in the way that scientists aim for.

1.4 Biomaterials and Tissue Engineering

Tissue engineering (TE), also known as tissue regeneration, is a growing biomedical discipline. Tissue engineering is a multidisciplinary branch of science that joins materials science, cell and molecular biology, biochemistry, engineering, medical implant science, and transplantation science to fabricate synthetic functional constructs that repair, maintain, or improve damaged tissues or whole organs by a combination of biomaterials, engineering, and living cells [6, 7]. The applied form of biomaterial used in tissue engineering is scaffold which has an essential role in the tissue regeneration process. A scaffold is a substrate with a 3D interconnected porous structure that imparts the necessary support for cells to proliferate, differentiate, and deliver cell signals, and it provides a microenvironment of tissue formation, in addition, it enables the cells to get rid of their wastes thanks to this porous structure. In fact, the process of creating new tissue using the tissue engineering method is not easy and requires special care in every step, starting from designing the scaffold in terms of its construction and composition to dealing with living cells and providing the appropriate conditions for them so that they can proliferate and differentiate in the required manner. Figure 1.5 shows the steps of culturing cells (such as MSCs, stem cells, and epithelial cells) onto the scaffold to form new ones. Firstly, the scaffold is fabricated with appropriate architecture, mechanical properties, and chemical composition. The cells are isolated either directly from the patient or using multipotent stem cells, such as mesenchymal stem cells (MSCs). The cells are seeded on the scaffold and incubated in the culture media under suitable conditions to proliferate and differentiate to specific cell phenotypes. During this, the scaffold degrades with a rate of cell

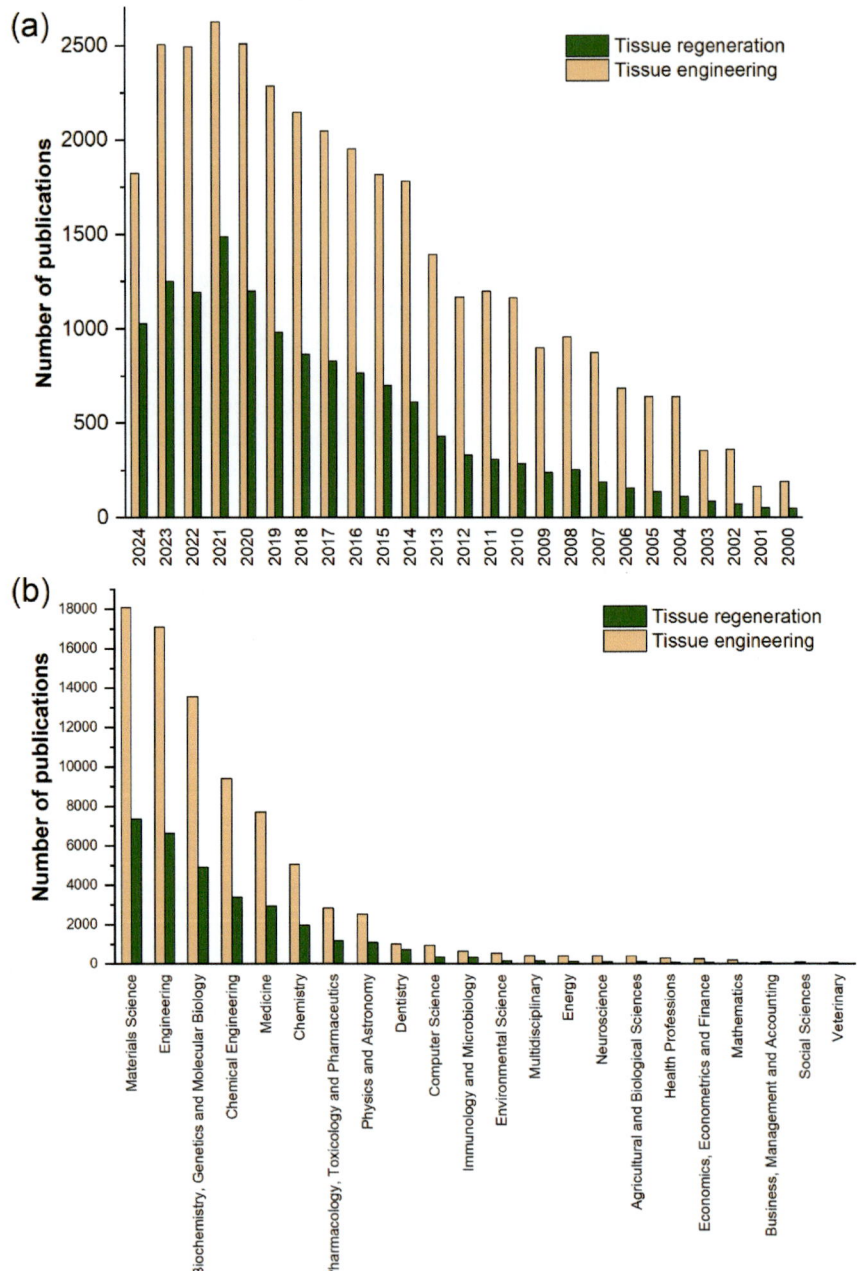

Fig. 1.4 a Number of published papers in the last 25 years, and **b** most relevant scientific field to tissue regeneration in the last 25 years. Search entities are (1) "biomaterials" and "tissue engineering, and (2) "biomaterials" and "tissue regeneration" according to the Scopus database

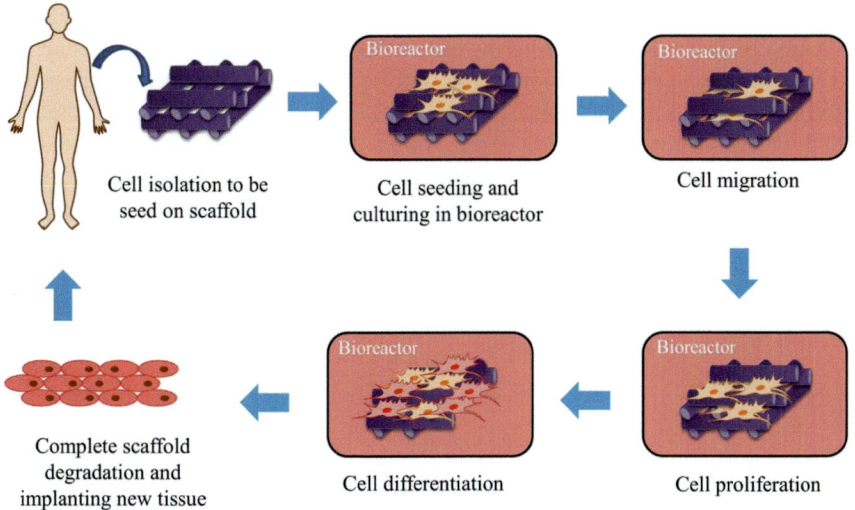

Fig. 1.5 Schematic diagram showing steps of formation of tissue by combination of scaffold and living cells

proliferation and differentiation rate, and finally, tissue is formed and ready to be implanted into the body. This method is called in vitro tissue engineering. There are other two types of tissue engineering; in vivo and in situ. In the case of in vivo tissue engineering the scaffold is incubated subcutaneously inside the body, under the physiological conditions in the body, new tissue is formed on the scaffold which can be collected and implanted in the required place. In situ tissue engineering is based on the implantation of the scaffold (with/without cells) directly into the diseased part, and it subsequently regenerates the lost tissue.

However, the fabrication of scaffolds possessing ideal properties is still a critical issue. That is because a typical scaffold biocompatibility depends on its physical properties (architecture, mechanical strength, wettability, roughness etc.), chemical composition, vascularization ability, nutrient delivery, cell metabolic waste removal, and cell penetration depth. The ideal scaffold should characterized by the following characteristics to perform its intended function and biological response [8]:

1. Highly 3D interconnected porous structure to achieve ideal microenvironment for cell migration, proliferation, differentiation, nutrient flow, and metabolic cell waste getting rid.
2. With controlled degradation and resorption rates that match cell growth.
3. Has mechanical properties that match the mechanical properties of implanted tissue.
4. Desirable surface roughness for cell attachment.
5. Suitable chemical composition is compatible with the cell.
6. Easy to be prepared in simple and complex shapes.
7. Similar to biological function and structure extracellular matrix (ECM).

Accordingly, the fabrication of an ideal scaffold depends mainly on two basics; materials chemistry and manufacturing techniques. The following chapters (Chaps. 2 and 3) will present types of biomaterials and techniques used in the synthesis of tissue engineering scaffolds. Also, ex vivo fabrication of specific tissue in biofactor including cells, proteins, and/or genes within the scaffold will be presented in Chap. 4 which demonstrates the tissue engineering applications in the regeneration of removed tissue or organ due to trauma or surgical interpolation. Moreover, modern trends and challenges of tissue reconstruction using the tissue engineering concept are presented in Chap. 5.

References

1. Montagnani C (1986) Pediatric surgery in Islamic medicine from the middle ages to the renaissance. In: Historical aspects of pediatric surgery. Springer, Berlin, pp 39–51
2. Saad MN (2016) Could Al-Zahrawi be considered a biomedical engineer? [Retrospectroscope]. IEEE Pulse 7:56–67
3. Marin E (2023) History of dental biomaterials: biocompatibility, durability and still open challenges. Heritage Sci 11:207
4. Aid M. https://www.muslimaid.org/get-involved/the-islamic-golden-age/al-zahrawi/
5. Porada CD, Atala AJ, Almeida-Porada G (2016) The hematopoietic system in the context of regenerative medicine. Methods 99:44–61
6. Vacanti JP, Langer R (1999) Tissue engineering: the design and fabrication of living replacement devices for surgical reconstruction and transplantation. Lancet 354:S32–S34
7. Williams D (2004) Benefit and risk in tissue engineering. Mater Today 7:24–29
8. Hutmacher DW (2001) Scaffold design and fabrication technologies for engineering tissues—state of the art and future perspectives. J Biomater Sci Polym Ed 12:107–124

Chapter 2
Types of Biomaterials Used for Tissue Engineering

Abstract As mentioned before, the response of living cells towards the scaffold material plays a vital role in the selection of material used in the fabrication of tissue engineering scaffold. Accordingly, the number of published articles concerning biomaterials used in tissue engineering applications has progressively increased. And so, the field of materials science, generally, and biomaterials, specifically, plays an important role in the development of tissue engineering. Therefore, most of materials have been studied and developed for scaffold fabrication. Due to the large number of these materials, it was approved to classify them. Several ways have been used to classify them; meanwhile, three ways for classification have been accepted by many materials scientists. The following sections discuss the classification of biomaterials and their applications. Biomaterials used in the fabrication of scaffolds, such as bioinert ceramic, bioactive ceramics, polymers, and composites, are discussed herein, however, metal biomaterials are not presented here because their applications in regenerative medicine are limited as a result of high-temperature fabrication of metal scaffolds which is unsuitable for in situ cell and biomolecules incorporation.

2.1 Classification of Biomaterials

Biomaterials can be classified based on three ways (Fig. 2.1): interaction with the surrounding tissues (inert, bioactive, and bioresorbable), the chemical structure of biomaterials (polymer, ceramic, metal, and composites), finally their origin (natural or synthetic). Bioinert materials (such as alumina and metals) cannot form bonds with living tissues, and they are isolated by fibrous connective tissue from the implanted tissues [1]. Bioactive materials (such as 45S5 glass) are the materials that can form bonds with living tissues [2]. Bioabsorbable materials (such as chitosan natural polymer) are materials that degrade at the rate of formation of new tissue [3]. There is an intersection among these ways of classification. For example, bioglass is a bioactive ceramic material, and it can be used as a tissue regenerative material. And so, the following sections present the type of materials and focus on their origin, chemical structure, and their tissue interaction.

M. M. Farag and Z. M. Al-Rashidy, *Biomaterials for Tissue Regeneration*, SpringerBriefs in Materials, https://doi.org/10.1007/978-3-031-75754-9_2

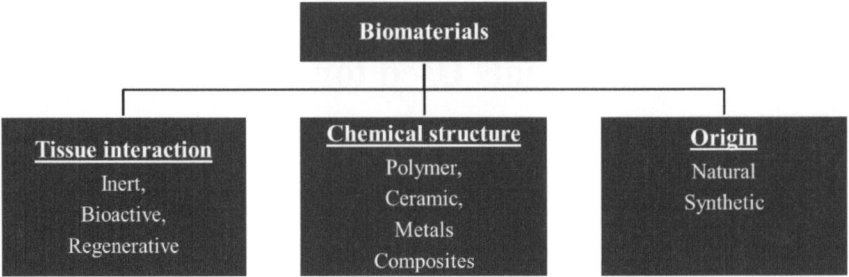

Fig. 2.1 Classification of biomaterials based on tissue interaction, chemical structure, and origin

2.2 Ceramic Biomaterials

Ceramic is neither metallic nor organic material composed of inorganic metallic and non-metallic compounds bonded by ionic or covalent bonds. It can be crystalline, glassy, or a mixture of both [4]. The origin of word "ceramic" comes from The Greek word "keramikò," which implies "burnt stuff" [5], where the ceramic is usually made by firing clays, and it is one of the oldest materials known by humankind. Ceramic materials, including glasses, refractories, cements, abrasives, or advanced ceramics, are widely used in electronic, optical, energy-related, nuclear, and biomedical applications. Ceramic materials are characterized by high mechanical strength, high chemical durability, resistance to high temperature, and high insulation properties, but they are brittle and fragile materials. Ceramics used in biomedical applications are known as bioceramics. From the side of tissue interaction, there are three types of bioceramics; inert (e.g. alumina), bioresorbable (e.g. Mg-phosphates), and bioactive (e.g. 45S5 glass) ceramic.

2.2.1 Bioinert Ceramics

As mentioned before, bioinert ceramic cannot form a chemical bond with the host tissue when it is implanted into the body. Alumina and zirconia have been the most widely used bioinert ceramics for a long time. They have been applied for tooth implants and hip prostheses. They are characterized by their inertness and low wear rate compared to metals and polymers. Figure 2.2 shows some examples of biomedical products made of alumina and zirconia already present in the market. More wear, abrasion resistance, and high mechanical properties could be obtained by a combination of alumina and zirconia which is known as alumina toughened zirconia (ATZ). It is a composite of small alumina particles dispersed in a very fine zirconia particle matrix. It has been used for hip prostheses due to its long lifetime. Maji and Choubey prepared ATZ by adding different percentages (0–30%) of alumina to 3 mol.% yttria-stabilized tetragonal zirconia polycrystals and they

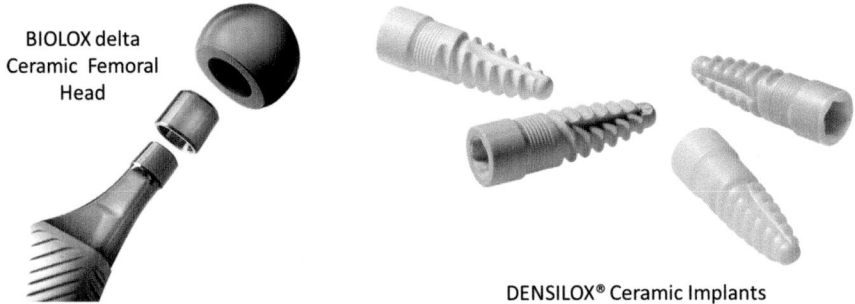

Fig. 2.2 Examples of alumina and zirconia are present in the market. BIOLOX delta is an alumina ceramic femoral head, while DENSILOX® is a zirconia dental implant

measured mechanical strength and defined formed crystalline phases. The results showed that as alumina increased, fracture toughness and porosity decreased, while hardness and bulk density increased. Accordingly, ATZ properties can be tailored in this study to be applied for specific biomedical applications [6]. Meanwhile, we are not focusing here on bioinert materials because their tissue engineering applications are limited.

2.2.2 Bioactive Ceramics

Bioactive ceramics are class materials that can make chemical bonds with the surrounding tissue, and they are reported to show better tissue responses than metals and polymers. They include calcium phosphate ceramics (e.g. hydroxyapatite trical-cium phosphate), bioactive glasses (e.g. 45S5 and 58S glasses), and glass–ceramics (e.g. apatite-wollastonite).

2.2.2.1 Calcium Phosphate Ceramics

Calcium phosphate ceramics (CaPs) are characterized by excellent biocompatibility due to the major constituent of bone is Ca-phosphate. Several types of CaPs depend on Ca/P ratio (ranges from 0.5 to 2), as shown in Table 2.1. The biodegradation, and hence biocompatibility of CaPs depends on Ca/P ratio. When Ca/P ratio < 1, CaPs possess a high dissolution rate which causes trouble during biological implantation. Whereas, a severe decrease in the degradation occurs when Ca/P is more than 1.67 severely decreases [7]. The solubility of Ca-phosphate members is arranged from high solubility to low one as follows [8];

$$ACP \rightarrow \alpha\text{-}TCP \rightarrow \beta\text{-}TCP \rightarrow CDHAp \rightarrow HAp \rightarrow fluorapatite$$

Table 2.1 Different types of CaPs compounds vary in Ca/P ratio [9, 10]

Name	Abbreviation	Formula	Ca/P
Monocalcium phosphate monohydrate	MCPM	$Ca(H_2PO_4)_2 \cdot H_2O$	0.5
Monocalcium phosphate anhydrous	MCPA	$Ca(H_2PO_4)_2$	0.5
Dicalcium phosphate dihydrate (brushite)	DCPD	$CaHPO_4 \cdot 2H_2O$	1
Dicalcium phosphate anhydrous (monetite)	DCPA	$CaHPO_4$	1
Octacalcium phosphate	OCP	$Ca_8(HPO_4)_2(PO_4)_4 \cdot 5H_2O$	1.33
α-Tricalcium phosphate	α-TCP	$Ca_3(PO_4)_2$	1.5
β-Tricalcium phosphate	β-TCP	$Ca_3(PO_4)_2$	1.5
Amorphous calcium phosphate	ACP	$Ca_x(PO_4)_y \cdot nH_2O$	1.2–2.2
Calcium deficient hydroxyapatite	CDHAp	$Ca_{10-x}(HPO_4)_x(PO_4)_{6-x}(OH)_{2-x}$ $(0< x< 2)$	1.5–1.67
Hydroxyapatite	HAp	$Ca_{10}(PO_4)_6(OH)_2$	1.67
Oxyapatite	OXA	$Ca_{10}(PO_4)_6O$	1.67
Tetracalcium phosphate (hilgenstockite)	TetCP	$Ca_4(PO_4)_6O$	2

Hydroxyapatite, HAp (formula, $Ca_{10}(PO_4)_6(OH)_2$) is the most commonly used CaP because it resembles the inorganic constituent of teeth and bone. The crystal symmetry is monoclinic (P21/b) symmetry and hexagonal (P63/m) symmetry which resembles HAp of natural bone [11]. Tricalcium phosphates, TCP (α- and β-forms) come in second place after HAp in terms of widespread use. Ca/P ratio of HAp and TCPs are 1.5 and 1.67, respectively, accordingly, the dissolution rate of TCP is more than that of HAp. And so, β-TCP is usually combined with HAp to increase its dissolution rate. Ca-phosphates have been prepared by several routes. Synthesis routes are presented in Fig. 2.3.

Synthesis Routes of CaP Ceramics

As shown in Fig. 2.3, synthesis methods of CaP ceramics are categorized into 4 main groups; dry synthesis, wet chemical synthesis, high temperature synthesis, and biogenic source synthesis. Herein, we'll present the most applied methods used for CaPs synthesis.

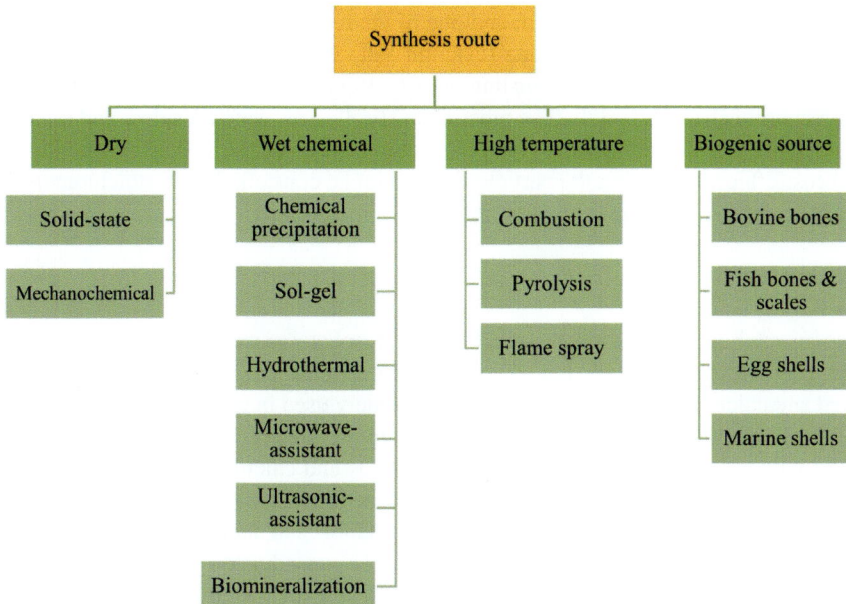

Fig. 2.3 Schematic presentation of different synthesis routes of CaPs

Solid-State Method

The synthesis of CaP in the solid state method is carried out without a solvent medium, and the precursors are mainly powder. Where the starting materials are mixed well by ball mill and then sintered at a high temperature (> 1000 °C). The product is characterized by high crystallinity, but there is a variety of the formed phases, large particle size distribution is predominant, and it is difficult to produce nanoparticle ceramic powder. This method is suitable for the preparation of HAp, HAp/TCP, and ion-doped β-TCP [12].

Mechanochemical Method

The mechanochemical method is another dry method which appropriate for mass production. The precursors with stoichiometric ratio are ground in a planetary ball mill in a solvent medium. The final product is affected by medium type, ball mill speed, duration, and type of starting materials. The advantages of this method are; the capability to prepare nanoparticles, the homogeneity of formed phases, and it is a desirable method for mass production of fluorapatites [13].

Chemical Precipitation Method

The chemical precipitation method is one of the wet chemical methods. It is the most widely method used for the synthesis of CaPs, specifically HAp. In this route, water-soluble precursors are used, and one of the reactants (usually phosphate salt) is added dropwise on another one (calcium salt) with stirring. The pH of mixed reactants is increased to precipitate immature CaP crystals. Then, the solution is aged to precipitate all CaP crystals. The precipitate is collected, washed, dried, and finally sintered at a suitable temperature [14]. This method is characterized by its simplicity and cost-effectiveness, but the particle size cannot be controlled, non-stoichiometric phases can be formed, and the particles are less crystallized.

Sol–Gel Method

Sol–gel method is based on the reaction of phosphate source and calcium source in an organic solvent, and then the reactants condense to form a gel. The gel is dried and calcined thereafter. This method is widely used in coating metal implants, such as titanium and magnesium alloys. For example, Catauro et al., coated titanium metal with sol–gel-derived HAp. They used P_2O_5 and calcium nitrate tetrahydrate $(Ca(NO_3)_2 . 4H_2O)$ as a precursor for phosphorus and calcium, respectively. Firstly, the precursors were dissolved separately in ethanol, and then phosphate solution was added dropwise to calcium solution, and pH was adjusted to 11 by ammonia solution. The above solution was aged for 1 day and then used for coating of titanium metal by dip coating method [15]. Sol–gel method is characterized by its good homogeneity, narrow particle size distribution, low synthesis temperature, and it can be used for coating applications, but the main drawbacks of this method are high cost and utilization of organic solvents which are unviable for the environment.

Hydrothermal Method

The hydrothermal method carries out the reaction of starting materials in an aqueous medium under high temperature and pressure. The physical properties of the final product are essentially affected by pH and temperature [16]. The advantage of the hydrothermal method is the possibility to prepare well-crystallized CaP ceramics with stoichiometric composition. But, the main disadvantage is the difficulty in regulating the particle size and morphology. Meanwhile, this disadvantage can be avoided by using either chelating agents or organic surfactants [17].

High-Temperature Methods

High-temperature methods, including pyrolysis, combustion, and flame spray, utilize high temperatures to partially burn or completely burn the precursors. The combustion process differs from the pyrolysis process. In combustion, the precursors burn in atmospheric oxygen, while the precursors are exposed to very high temperatures under an inert atmosphere. Fuels are usually used in high-temperature methods, and organic fuels, such as sucrose, glycine, urea, and citric acid are considered suitable fuels for this reaction. The main factors controlling this method are temperature and precursor concentrations [18]. This method is characterized by high homogeneity of the chemical composition of the end product, highly dense particles that increase mechanical strength, and very small particle size resulting from rapid cooling. However, the essential drawback of this method are formation of agglomerates which can be diminished by attrition.

Biogenic Source Method

Biogenic sources, including fish bones and scales, coral, sea-shell, eggshell, and bovine bone, used for the synthesis of CaPs are drawn the interest due to their cost-effectiveness, and eco-friendly and the end product resembles the apatitic bone body [19]. However, impurities present in biogenic sources almost cause nonstoichiometry in the composition which is considered the essential drawback of this method. Coral, sea-sell, and eggshells are biogenic sources of calcium carbonates which can react with phosphorous sources to form CaPs. Animal bone can be converted directly to CaP by thermal decomposition, or the bone is deproteinized and defatted first, and then calcined. Synthesis of CaPs from biogenic calcium carbonate sources is more expensive than animal bone sources, that is because the required heat treatment of calcium carbonate to convert it to calcium oxide which usually reacts with phosphoric acid to form CaP, while animal bone is directly transformed to CaP because it contains calcium and phosphorous.

Calcium phosphate cements (CPCs) have been discovered since 1983. From this time, there have been many formulations of CPCs developed and proposed. However, in recent times, CPCs are based on brushite and HAp (or Ca-deficient HAp) as end products of cement reaction. Brushite is more soluble and resorbed faster than HAp. The most common formulations of CPCs are; brushite, TeCP-based apatite, and α-TCP-based apatite. Cement reaction of the first two types occurs by acid–base reaction, while the last one occurs by a hydrolysis reaction, and all CPCs set by dissolution and precipitation process in which it occurs in three stages: reactant dissolution, new phase nucleation, and finally crystal growth. Brushite CPC is produced from a reaction of monocalcium phosphate monohydrate (MCPM) with β-TCP according to Eq. 2.1. The reaction setting time is relatively short. TeCP-based apatite CPC results from a reaction of TeCP with dicalcium phosphate anhydrous (DCPA) which gives HAp following Eq. 2.2. The setting time of this cement reaction is so long. The last α-TCP-based apatite CPC happens by hydrolysis of 3α-$Ca_3(PO_4)_2$ by water to give calcium deficient HAp concerning Eq. 2.3 [20]. Meanwhile, the ultimate properties of cement are generally determined by the particle size of the starting powder and the liquid-to-powder (L/P) ratio.

$$Ca(H_2PO_4)_2 \cdot H_2O + \beta\text{-}Ca_3(PO_4)_2 + 7H_2O \rightarrow 4CaHPO_4 \cdot 2H_2O \qquad (2.1)$$

$$2Ca_4(PO_4)_2O + 2CaHPO_4 \rightarrow Ca_{10}(PO_4)_6(OH)_2 \qquad (2.2)$$

$$3\alpha\text{-}Ca_3(PO_4)_2 + H_2O = Ca_9(HPO_4)(PO_4)_5OH \qquad (2.3)$$

2.2.2.2   Magnesium Phosphates

Magnesium phosphate (MgP) ceramics have possessed wide application in agricultural and civil engineering fields, and lately, they have shown promising biomedical applications. Magnesium is the fourth most abundant cation in the human body [21]. Magnesium phosphate is vital in the formation of bones, teeth, and muscles, and its benefits in the metabolism of fats, proteins, and carbohydrates. Accordingly, magnesium phosphate MgP has been demonstrated in increasing demand in orthopedic and dentistry. In certain cases, they are desirable to alternate the aforementioned Ca-phosphate bone substitutes. That is because of their higher solubility than Ca-phosphates inside the body, as well as, their derived cements possess higher mechanical properties than well-known bone substitutes, like brushite [22, 23]. Like Ca-phosphates, there are different types of Mg-phosphate compounds with different Mg/P ratios which are shown in Table 2.2, but the maximum limit of Mg/P ratio is 1.5. On the other hand, the majority of the Mg-phosphate compounds are hydrated at low temperatures, while most Ca-phosphate compounds do not possess crystallization water in their structures.

Mg-phosphate cements (MPCs) have been used for a long time in civil engineering due to their high early mechanical strength [24]. Recently, MPCs became important ceramic cement applied in orthopedics. The cement is mainly based on struvite ($NH_4MgPO_4 \cdot 6H_2O$) and K-struvite ($KMgPO_4 \cdot 6H_2O$) crystals [25–27]. In contrast to CPCs, MPCs are characterized by high mechanical strength and degradation rates [23]. Moreover, MPCs stimulate osteoblast differentiation and they present good biocompatibility with the bone cells [28, 29]. In addition, some MPCs contained sodium ions which showed antibacterial activity [30, 31].

Cementation of MPCs occurs by acid–base reaction. The suggested mechanism of MPCs was presented by Soudée and Péra based on MgO and $NH_4H_2PO_4$ (ammonium dihydrogen phosphate) [32]. Firstly, MgO is hydrated by 6 water molecules to

Table 2.2   Known Mg-phosphates with chemical formula and Mg/P ratio [36]

MgP compound	Chemical formula	Mg/P
Bobierrite	$Mg_3(PO_4)_2 \cdot 8H_2O$	1.5
Brucite	$Mg(OH)_2$	–
Cattiite[a]	$Mg_3(PO_4)_2 \cdot 22H_2O$	1.5
Dittmarite	$NH_4MgPO_4 \cdot H_2O$	1
Farringtonite	$Mg_3(PO_4)_2$	1.5
Hannayite	$(NH_4)_2Mg_3(HPO_4)_4 \cdot 8H_2O$	0.75
Magnesia	MgO	–
Newberyite	$MgHPO_4 \cdot 3H_2O$	1
Schertelite	$(NH_4)_2Mg(HPO_4)_2 \cdot 4H_2O$	0.5
Struvite	$NH_4MgPO_4 \cdot 6H_2O$	1
K-struvite	$KMgPO_4 \cdot 6H_2O$	1

[a]Metastable Mg-phosphate in water

form $Mg(H_2O)_6{}^{2+}$ octahedron complex as shown in Fig. 2.4. Each octahedron held to one phosphate tetrahedron ($PO_4{}^{3-}$) and one $NH_4{}^+$ ion by hydrogen bonds to form a struvite crystal ($NH_4MgPO_4 \cdot 6H_2O$). K-struvite ($KMgPO_4 \cdot 6H_2O$) crystals present as the main product of MPC when ammonium dihydrogen phosphate is replaced by potassium dihydrogen phosphate (KH_2PO_4). However, the cementation reaction of MPCs is exothermic and usually fast reaction which requires retarders to decrease the initial setting time and evolved temperature. Common retarders used in MPCs are glacial, acetic acid, borax, boric acid, and sodium triphosphate. The concept of MPC chemistry was applied in our previous works to prepare MgP-based scaffolds at room temperature. Where we fabricated bone regeneration and antibacterial MgP scaffolds loaded with lysozyme drug using paste extruding deposition rapid prototyping technique, and the scaffold green bodies were post-hardened by ammonium phosphate solution. The scaffolds are characterized by well-interconnected porous structures, good biodegradability, mechanical strength, and cell viability [33]. In a prolonged work, we synthesized MgP/gelatin scaffolds using the same previous protocol. Herein, gelatin was used with different percentages to control scaffold degradation, porosity, drug release, and mechanical. Moreover, all the scaffolds showed good cell viability [34].

CPCs have been added to MPCs to combine the advantages of both compounds. For example, not limited to, newberyite and/or MgO combined with Ca-phosphates, such as MCPM, DCPA, DCPD, TCP, and TTCP, to obtain promising cement with good biocompatibility, biodegradation, practical setting time, and mechanical properties [35].

2.2.2.3 Bioactive Glass and Bioactive Glass–Ceramics

Bioactive glass and glass–ceramic biomaterials have been extensively used in orthopedic, dental, and tissue engineering, and there already are numerous products based on them present in the market as shown in Table 2.3 [37]. Glass is defined according to the American Society as an inorganic product of fusion which has cooled to a rigid condition without crystallizing [38]. It is obtained by the melting of mixed oxides of glass constituents at high temperatures followed by sudden cooling of the melt in air or water. While glass–ceramic is a polycrystalline material with residual glass obtained from controlled heat treatment of specific glass [39]. And so, a poorly ordered or random amorphous structure of glass is transformed into a well-ordered structure by the crystallization process as shown in Fig. 2.5.

Bioactive glass (BG) and bioactive glass–ceramic are a type of biomaterials that have drawn attention during the last decades in the medical field for the treatment of hard and soft tissues. Since the discovery of BG, the prevailing concept of biological materials has changed, and it led to the discovery of a second generation of biomaterials. Where BG can form a bond with the host tissue via the formation of HAp layer on its surface after immersion in the physiological fluid, and so it promotes osteogenesis and angiogenesis [40–42]. Hench and coworkers discovered the first type of BG in the late 1960s, they prepared it by conventional melting method of silicate-based

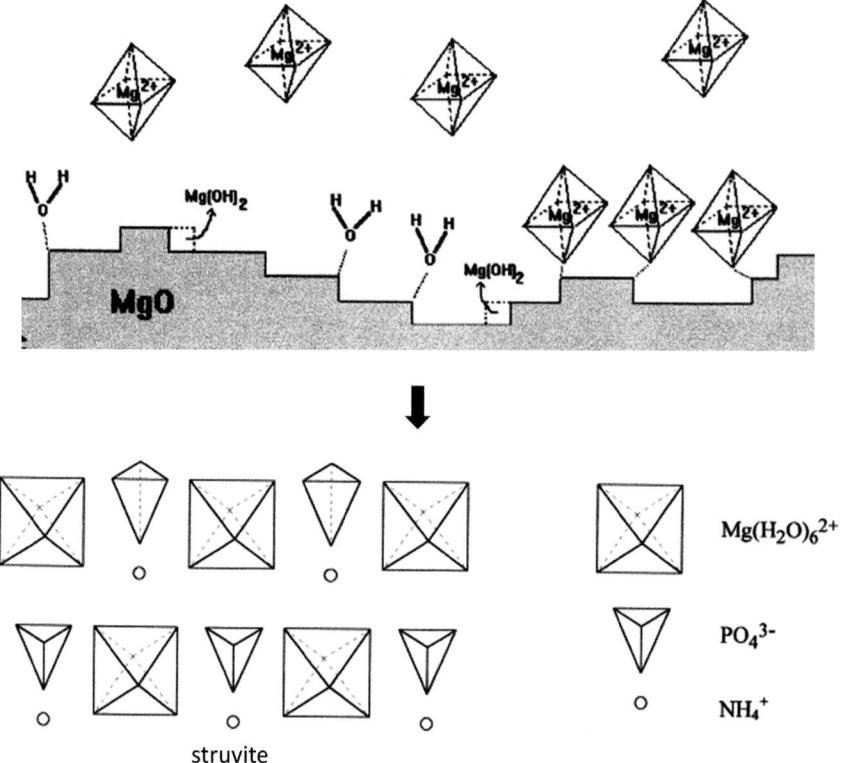

Fig. 2.4 Mechanism of cement reaction of MPCs based on MgO and $NH_4H_2PO_4$. After modification of [32]. Copyright Elsevier 2000

glass. They examined numerous glass compositions based on the quaternary system of SiO_2–CaO–Na_2O–P_2O_5 by fixing P_2O_5 constant (with 6 wt%), and consequently construed a diagram known as Hench's diagram which divided into four zones based on bioactivity index (Fig. 2.6). Bioactivity Index (IB) is the time required for more than half of the interface to bond ($t_{0.5bb}$) as in Eq. 2.4.

$$IB = 100/t_{0.5bb} \qquad (2.4)$$

Zone A is called the bioactive zone, in which the glass forms chemical bonds with the host tissue in less than 1 month. Zone B is a bioinert zone, where the glass reactivity is too low reactivity causing the formation of fibrous tissues encapsulates the glass. Zone C is a resorbable zone, but herein the resorbability is too much and the glass dissolves quickly before 1 month. Zone D is a non-glass formation zone, in which the glass is devitrified [43]. In zone A Hench and coworkers determined glass based on 45.0 SiO_2–24.5 CaO–24.5 Na_2O–6.0 P_2O_5 wt% and named it later as 45S5

Table 2.3 Different bioactive glass and glass–ceramic products present in the market

Product name	Composition	Applications
MEP®	45S5	Replacement of ossicles in the inner ear
Bioglass-EPI®	45S5	Coating to anchor cochlear implants to the temporal bone
ERMI®	45S5	Encourages bone formation after tooth extraction and acts as roots for dental implants
PerioGlas®	45S5	Periodontal disease
Biogran®	45S5	Bone defects in the jaw
ArGlaes®	Ag-doped BG	Antibacterial wound healing and burn care
TheraSphere™	Y_2O_3–Al_2O_3–SiO_2	Radiology
NovaBone®	45S5	Orthopedics non-bearing bone graft
UniGraft®	Silicate	Antibacterial teeth filler
Medpor®-Plus	45S5/polyethylene composite	Ocular bone graft
NovaMin®	45S5	Toothpaste as a hypersensitivity inhibitor, treatment of gingivitis
BoneAlive®	S53P4	Bone infections, diabetic foot osteomyelitis, spinal fusions
Glassbone®	45S5	Orthopedic, grafting sockets for dental implants
Cortoss®	45S5/bisphenol-a-glycidyl and bisphenol-a-ethoxy methacrylate composite	Bone cement
Vitoss® BA	45S5 (porous)	Orthopedic
Signify®	45S5/polyethylene glycol/ glycerol composite	Bone defect filler
Glace™	S53P4	Cranial/Maxillofacial
NanoFuse®	45S5 + demineralized bone	Orthopedic
DermaFuse or Mirragen	13-93B3	Wound healing
Activa™	Glass ionomer	Dental cement
OssiMend® Bioactive	45S5/type 1 collagen/carbonate apatite	Orthopedic
BioMin® C and F	Fluorine or chlorine-contained silicate	Dental
SIGNAFUSE	45S5 + (60% HAp and 40% TCP)	Orthopedic
Dicor®	Mica-based glass–ceramic	Dental (veneers, inlays, onlays, and dental crowns)
Ceramco®	Leucite-based glass–ceramic	Dental

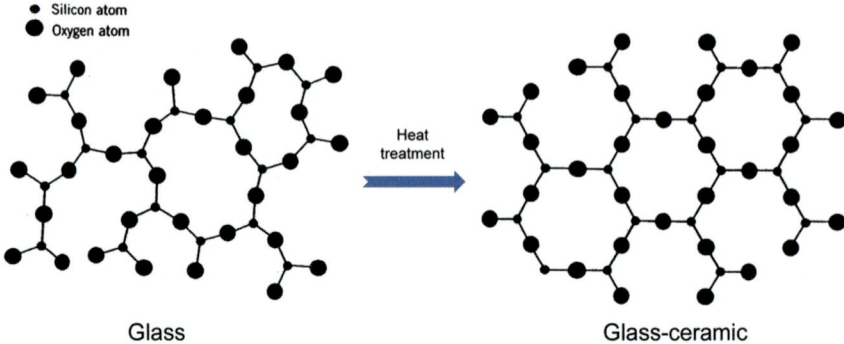

Fig. 2.5 Difference between crystalline and amorphous silica is an example to show the difference between the glassy and crystalline states

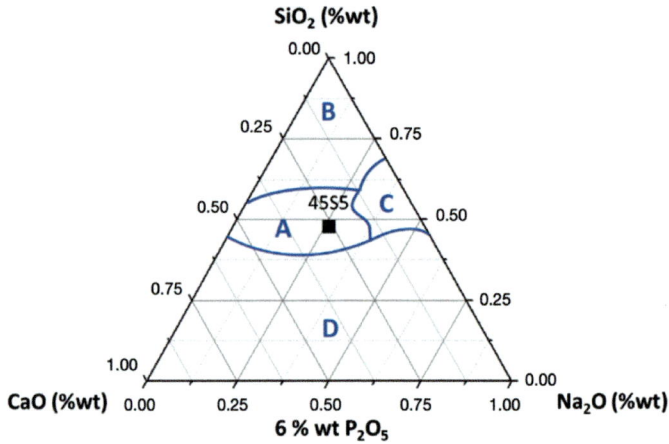

Fig. 2.6 Hench diagram based on SiO_2–CaO–Na_2O–P_2O_5 by fixing P_2O_5 constant (with 6 wt%), and it divided into four zones depending on bioactivity index [43]. MDPI, This article is an open access licensed under a Creative Commons Attribution 4.0 International License

Bioglass® as the best bioactive glass in their diagram [40]. 45 means the weight of SiO_2, and 5 is Ca/P atomic weight ratio in the glass.

Conversion of the glass surface to HAp at the glass/tissue interface is carried out following a mechanism described by Hench [40, 44, 45] and others [46, 47]. When the glass is implanted in the body, HAp is formed in the following successive steps (Fig. 2.7):

(1) After implantation in the body, exchange occurs instantly between Na^+ and/or Ca^{2+} in the glass with H^+ or H_3O^+ in the body fluid according to Eq. 2.5.

$$Si-O-Na^+ + H^+ + OH^- \rightarrow Si-OH + Na^+_{(solution)} + OH^- \qquad (2.5)$$

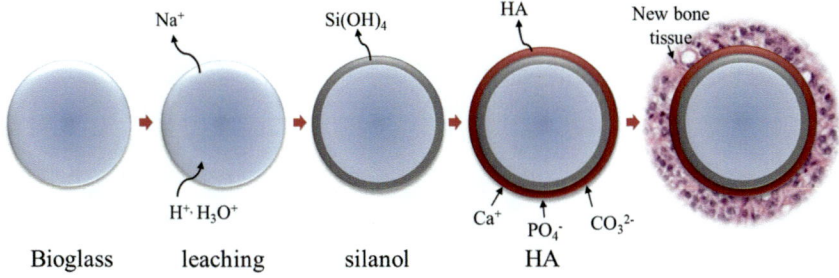

Na$^+$ Si(OH)$_4$ HA   New bone tissue

H$^+$· H$_3$O$^+$ Ca$^+$ PO$_4^-$ CO$_3^{2-}$

Bioglass   leaching   silanol   HA

Fig. 2.7 Schematic presentation of the steps of HAp formation on the glass surface after contact with the body fluid [48]. Springer Nature 2023. This article is an open access licensed under a Creative Commons Attribution 4.0 International License

(2) Si–O bonds in the glass network are attacked by OH$^-$ ions to compensate the shortage of positive charges of leached cations (Na$^+$ and/or Ca^{2+}) leading to the breaking of Si–O–Si bridges, and subsequently silanol groups (Si–OH) are formed according to Eq. 2.6.

$$2(Si-O-Si) + 2(OH) \rightarrow Si-OH + HO-Si \qquad (2.6)$$

(3) Silanol groups condense and repolymerize on the glass surface to form a silica-rich layer (Eq. 2.7).

$$2(Si-OH) + 2(HO-Si) \rightarrow -Si-O-Si-O-Si-O-Si-O- \qquad (2.7)$$

(4) Ca^{2+} and PO$_4^{3-}$ migrate from the surrounding fluid to a silica-rich layer to form an amorphous CaP-rich layer on the glass surface which acts as nucleation sites for crystallization of HAp thereafter.

(5) OH$^-$ and CO$_3^{2-}$ are incorporated from the surrounding fluid to crystallize together with amorphous Ca-P to form hydroxyl carbonated apatite (HCA).

After the discovery of 45S5 bioactive glass, many other types of glass have been discovered based on silicate, borate, and phosphate systems. Borate bioactive glasses are characterized by low chemical durability and fast conversion to HAp compared to silicate-based bioactive glasses, and they induce cell differentiation proliferation, and tissue regeneration [49]. Unlike silicate BG, borate BG is wholly converted to HAp, while, HAp is developed only on the surface of silicate glass [50–52]. The mechanism of conversion of borate glass to HAp is nearly similar to that of silicate glasses. Firstly, Na$^+$ and Ca^{2+} cations are leached out from the glass surface leading to the breaking of B–O bonds resulting in the release of BO$_3^{3-}$ groups. In parallel, calcium ions on the glass surface attract phosphate ions from the solution to form HAp crystals ultimately. The conversion process continues underneath the newly formed HAp layer by penetrating phosphate ions through HAp tunnels present in HAp layer to attach to calcium ions and form a new layer of HAp, and the process

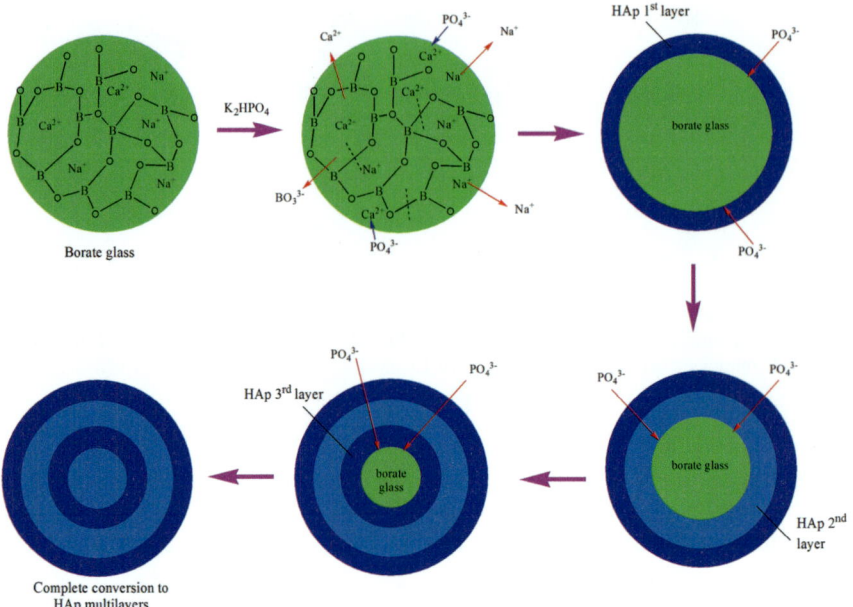

Fig. 2.8 Conversion mechanism of borate bioactive glass to HAp after immersion in phosphate solution [53]. Copyrights John Wiley and Sons, 2020

remains until conversion of the total glass body to HAp in a form of layers as shown in Fig. 2.8 [52, 53].

In addition, phosphate glasses, essentially based on $Na_2O–CaO–P_2O_5$ composition, are more favorable than Ca-phosphate ceramics in certain cases as bioresorbable materials. That is because of the flexibility to change their composition to tailor their dissolution rate, as well as, it is possible to add other metal oxides without committing to specific proportions to get functionalized glasses used for the treatment of specific diseases [42, 54].

Thanks to the discovery of sol–gel technology, sol–gel-derived BG was first prepared in 1991 using this method which produced BG with superior properties over melt-derived BG [55, 56]. Despite the melting method is cheap, simple, and solvent-free, BG prepared by sol–gel approach is distinguished by high surface area and porosity, high homogeneity, energy saving, a wide range of compositions can be obtained, can be used in the coating applications, and particles can be obtained in a nanoscale, and so nanofibers can be manufactured [57]. However, the sol–gel method is an expensive, complex, long synthesis process, and it needs mostly toxic solvents in the synthesis process.

Sol–gel process uses oxide solution precursors (usually alkoxides which are metal oxide attached with alkyl groups) and polymerizes them by conversion from liquid to a "sol" to a three-dimensional network called a "gel" [58]. Accordingly, the first step occurs in sol–gel is hydrolysis of metal alkoxide, and the second one is condensation. Hydrolysis is catalyzed by acid or base, and so it is known as acid and base catalysis, respectively. The factors controlling the sol–gel process are; alkyl group nature (number of carbons in the alkyl chain), type and concentration of catalyst used in the hydrolysis step, and amount of water. To explain the process of sol–gel reactions, silicon alkoxide is taken as an example of a metal alkoxide precursor. The hydrolysis step arises by the replacement of alkyl groups by hydroxyl groups as shown in Eq. 2.8. So, silicon alkoxide requires a minimum of 4 mol of water, and an increase in water content in the reaction increases the rate of hydrolysis [59]. However, silicon alkoxide is immiscible with water, therefore another solvent, such as alcohols, miscible with alkoxide is preferred to be used in the reaction. After complete hydrolysis of alkoxide, the condensation step starts with the polymerization of silicon ions in three-dimension by bridging oxygens via loss of water according to Eq. 2.9. Accordingly, a high degree of hydrolysis gives a highly branched silicate network as shown in Fig. 2.9 represents the progress of condensation of silanol groups and the formation of 3D interconnected glass network. As a result of continuous condensation steps, highly branched and agglomerates with small sizes formed in the solution are crosslinked to form the gel.

$$Si(OR)_4 + 4H_2O \rightarrow Si(OR)_4(OH)_4 + 4ROH \qquad (2.8)$$

$$Si-OH + HO-Si \rightarrow Si-O-Si + H_2O \qquad (2.9)$$

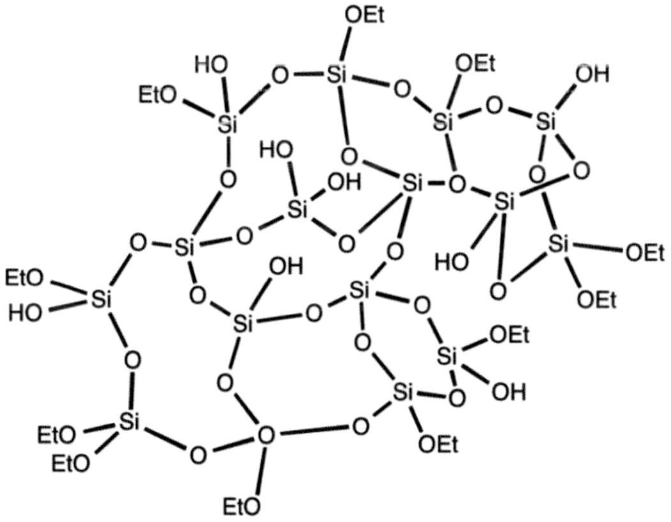

Fig. 2.9 Presentation of condensation of hydrolyzed silicon atoms (Si–OH) in 3D [48]. Copyrights Springer Nature 2023. This article is an open access licensed under a Creative Commons Attribution 4.0 International License

Table 2.4 Different therapeutic ions and their function incorporated in BG and bioceramics

Ion	Function and application
Bismuth	Antimicrobial, anti-inflammatory, treatment of gastrointestinal disorders and peptic ulcer
Chromium	Anticancer, antibacterial, insulin signaling, diabetes mellitus
Cobalt	Improves osteogenesis, antibacterial activity, anti-Inflammatory, anticoagulating agent
Copper	Improves angiogenesis and osteogenesis, antibacterial
Gallium	Anticancer, antimicrobial, anti-inflammatory
Germanium	Anticancer, antioxidant
Iron	Important in the transportation of oxygen and nutrition, vessel formation and tissue ingrowth Possess magnetic properties which can be used in cancer treatment by hyperthermia and drug delivery
Lithium	Treats mental disorders (e.g. depression and mania), promotes alkaline phosphatase and osteogenic gene expression in osteoblasts
Magnesium	Enhances bone formation in the early stages, essential in bone metabolism and bone mineralization process through controlling osteoblast and osteoclast activities
Manganese	Increase the osteoblast response and secretion of osteogenic substances
Potassium	Regulation of cellular electrolyte metabolism, nutrient transportation, cell signaling
Rare earths	Luminescent optics for imaging application
Silver	Antibacterial
Strontium	Osteogenic, angiogenic, antibacterial
Zinc	Antimicrobial, enhances angiogenesis and homeostasis

Bioactive glasses and bioceramics have been functionalized by therapeutic ions (such as Ag^+, Cu^{2+}, and Zn^{2+}) to increase their potential in biomedical applications. Different ions incorporated in BG and bioceramics and their application are in detail in [60] and [61], and summarized in Table 2.4.

2.2.3 Polymeric Biomaterials

Recently, polymeric biomaterials have been extensively used in different biomedical applications, specifically in tissue regeneration as biodegradable materials. That is because they can be applied to tailor degradation rate, mechanical properties, and surface area. Tailored degradation and mechanical properties are very important characteristics of tissue engineering scaffolds to be matched with the host tissue. The main factors affecting polymer biodegradation rate are crystallinity, hydrophilicity, and composition of the polymer. Biodegradable polymers can be divided, according to their origin, into a natural polymer (e.g. chitosan, collagen, gelatin, alginate, and cellulose) and synthetic polymer e.g. poly(ε-caprolactone) (PCL), poly(L-lactic acid) (PLLA), poly(glycolic acid) (PGA), poly(L-lactic-co-glycolic acid (PLGA), polyvinyl alcohol (PVA), etc.). Generally, natural polymers undergo enzymatic degradation, while synthetic polymers degrade by hydrolytic degradation.

2.2.3.1 Natural Polymers

Natural polymers can be extracted from plants, animals, and microorganisms. The distinctive advantages of natural polymers are biocompatibility, biodegradability, bioactivity, and they resemble the living tissues' extracellular matrix. While disadvantages of these polymers include low thermal stability, low mechanical properties, immunogenic response, presence of impurities, change of grade from batch to batch, and the complexities associated with their purification and risk of disease transmission. Overall, the advantages of natural polymers make them superior to other synthetic polymer in different biomedical applications. In this chapter, we'll show a brief of the most widely used natural polymers in tissue regeneration.

Collagen

Collagen is the most abundant natural polypeptide polymer in the human body, it is a fibrous protein present in skin, bone, muscles, cartilage, vessels, and veins. There are four main types of collagen, Type I, Type II, Type III, and Type IV, the most abundant one is Type I. Collagen is built of three chains that bind together through glycine to form a triple helix. Each chain is composed of three blocks of amino acids; glycine, proline, and hydroxyproline with percentages of 33%, 25%, and 25%, respectively [62] (Fig. 2.10). Collagen is extracted by chemical hydrolysis and enzymatic hydrolysis. Collagen is pretreated with acid or base to remove non-collagenous substances. In the chemical hydrolysis method, the pretreated collagen is dissolved in hydrochloric acid, acetic acid, or citric acid at 4 °C. Then, the supernatant is extracted and collagen is precipitated with NaCl and separated by filtration. Collected collagen redissolved in acetic acid and then dialyzed. The enzymatic hydrolysis process is similar to chemical hydrolysis, but enzymes, such as Flavourzyme®, pepsin, and Alcalase®, are added to the acid solution used to dissolve the pretreated collagen [63].

Gelatin

Gelatin is a polypeptide polymer produced from the controlled thermal hydrolysis of collagen. In which, covalent and hydrogen bonds of collagen have to break down by chemical treatment (acidic or alkaline treatment) before gelatin extraction. Accordingly, there are two types of gelatin based on the chemical pretreatment process which are; Type A and Type B corresponding acid and alkaline pre-treatment, respectively. Gelatin can be extracted from skin and bone animals (e.g. caws and porks) [64]. The

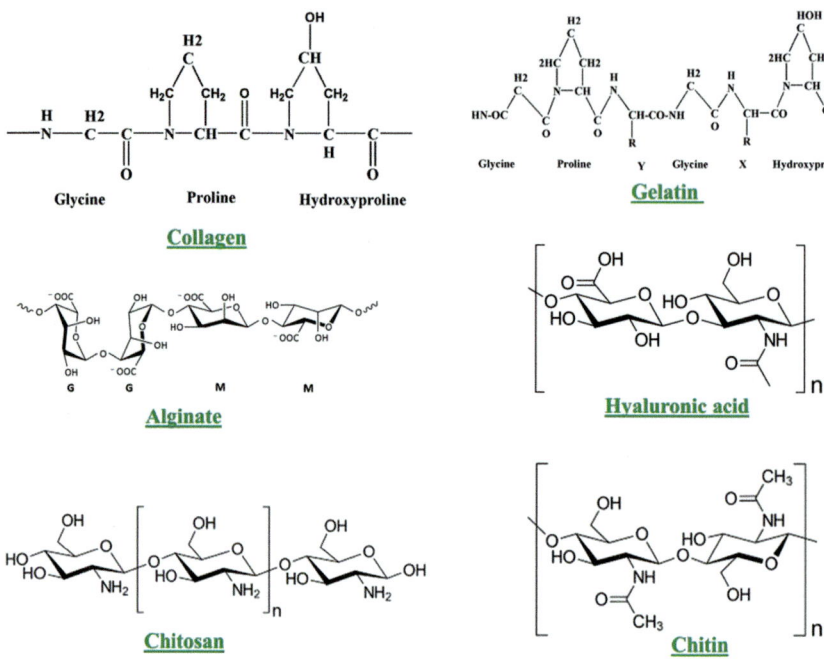

Fig. 2.10 Examples of structure of natural polymers; collagen, hyaluronic acid, chitosan, chitin, alginate, and gelatin

main advantages of gelatin are; low cost, excellent biocompatibility and biodegradability, availability, high viscosity, and it has no immunological response. In this respect, it has been used successfully in tissue engineering and drug delivery applications, specifically in the case of a combination of bioactive ceramics with it. However, gelatin possesses some drawbacks, such as a high degradation rate and low mechanical strength which can be solved by applying crosslinking or combining it with other polymer's low degradability and high mechanical strength (e.g. polycaprolactone) [65].

Fibrin

Fibrin is similar to collagen, it is derived from fibrinogen. It is composed of three major chains; central domain "E" with two peptide pairs A and B molecules, and two terminal domains of fibrinopeptide D amino acids. Fibrin is included in the blood clotting process through the polymerization of fibrinogen by serine protease thrombin and forms a hemostatic clot over a wound site which can be degraded by enzymes with a process known as the fibrinolysis process [66]. Therefore, fibrin sealant products have been derived from fibrin and applied potentially in tissue sealing after surgeries, such as intestinal, neural, and vascular procedures. In addition, platelets can be mixed with fibrin sealants to give platelet–fibrin gels used mainly in maxillofacial, dental, and orthopedic applications [67].

Hyaluronic Acid

Hyaluronic acid is a water-soluble linear polysaccharide polymer composed of N-acetyl-d-glucosamine and glucuronic acid alternating units. It is an important constituent in articular cartilage, skin, and endothelial cells, where, about 50% of hyaluronic acid of the human body is present in the skin, and it is produced during the early wound healing process. Moreover, hyaluronic acid promotes mesenchymal and epithelial cell migration and differentiation in tissue repair. Hyaluronic acid is synthesized by a fermentation process by microorganisms. The most widely used bacteria used for a high production rate is *Streptococcus zooepidemicus*, a Gram-positive bacteria [68].

Chitin and Chitosan

Chitosan is a linear polysaccharide produced directly by the deacetylation of chitin in the exoskeleton of an arthropod. There is an inverse relationship between chitosan degradation rate its degree of acetylation and crystallinity [69]. It is composed of N-acetyl D-glucosamine and D-glucosamine units [70] (Fig. 2.10). Distinctive properties of chitosan, such as antibacterial and antifungal activities, biodegradability, hemostatic activity, mucoadhesion, and cell viability, have drawn the interest to be used in tissue regeneration, drug delivery, orthopedics, and wound dressing. As well as, the structure of chitosan is similar to hyaluronic acid and glycosaminoglycans present in the human body and it undergoes enzymatic degradation by chitosanase, lysozyme, and papain enzyme. The mucoadhesive property of chitosan comes from its high positive charges on the polymer which interacts with the negatively charged mucous membrane. And so, it has been widely used as a mucoadhesive drug carrier [71]. However, like all biodegradable polymers, chitosan lacks bioactivity which limits its application in bone regeneration. Accordingly, bioactive ceramics have been introduced to improve the polymer bioactivity and formation of new bone tissue. Bioactive glass nanoparticles are excellent candidates as bioactive fillers for chitosan due to their high surface area, osteoconductivity, and osteoinductivity.

Alginate

Alginate is a polysaccharide binary copolymer composed of M blocks (β-D-mannuronic acid) and G blocks (C-5 epimer α-L-guluronic acid) (Fig. 2.10). It is extracted from brown algae (usually in sodium salt form), thereby, it is present in the cell walls and intercellular spaces of algae. Alginate is crosslinked by divalent cations, such as calcium ions to form a gel which makes it a desirable polymer for drug encapsulation [72]. In addition it has been widely used in wound dressing, injectable bone cement, and as bioink in scaffold prepared by 3D bioprinter. Despite cost-effective, low toxicity, excellent biocompatibility, and structure resembling the tissue extracellular matrices, alginate cannot undergo enzymatic degradation in the body, and so it possesses a low degradation rate at neutral pH. In this respect, the alginate degradation was modified by a physical approach, such as gamma irradiation and a chemical approach [73].

2.2.3.2 Synthetic Polymers

Synthetic polymers are usually bioinert materials. They are distinguished from natural polymers by their composition stability from batch to batch, high mechanical properties, their degradation properties can be tailored, and they almost undergo hydrolytic degradation which is characterized by minimal side effects that probably occur for patient variations compared to high probability of side effects happen by enzymatically degraded polymers [74]. Synthetic polymers are synthesized by polymerization (condensation) of low molecular weight monomer to high molecular weight polymer chain. Polymers are classified into three types; thermoplastics (they can be melted and remolded to the required shape), thermosets (they decompose upon heating rather than melting), and elastomers (or rubbers, which can be stretched and return to their original shape without deformation). We'll represent synthetic polymers widely used in different biomedical applications, and their chemical structures are presented in Fig. 2.11.

Poly(Lactic Acid) (PLA)

PLA is thermoplastic aliphatic polyester approved by the Food and Drug Administration (FDA). It possesses two optically active forms; D-lactide and L-lactide. It synthesized by ring opening polymerization process of L-lactide or D-lactide, and the type of final product of PLA depends on starting lactide type, and so there are poly(L-lactide), poly(D-lactide), and poly(D,L-lactide). The first two types are crystalline (about 37% crystallinity) and the third one is amorphous. Poly(L-lactide) has slow degradation, good tensile strength, high modulus, good biodegradability, and good biocompatibility. All these advantages make the use of this polymer successful in tissue engineering, drug delivery, and load bearing applications.

Polyglycolide (PGA) and Poly(Lactide-co-Glycolide) (PLGA)

Polyglycolide (PGA) and poly(lactide-co-glycolide) (PLGA) polymers are synthesized by ring-opening polymerization like PLA synthesis process. PGA has high crystallinity (45–55%), while PLGA is an amorphous polymer. PGA is synthesized

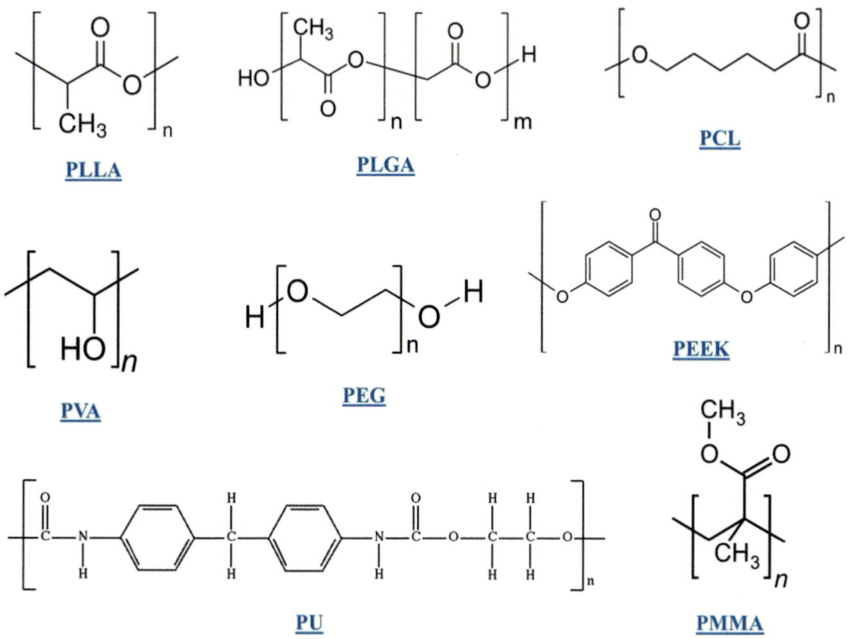

Fig. 2.11 Examples of structure of synthetic polymers; PLLA, PLGA, PCL, PVA, PEG, PU, PMMA, and PEEK

from glycolide, while PLGA is synthesized by co-polymerization of lactide and glycolide with different ratios, the degradation of the final polymer depends on the lactide:glycolide percentage. Thus, PLGA degradation can be ordered from high to low degradable as follows; 50:50 > 75:25 > 85:15 [75]. And so, the degradation of co-polymer can be tailored by changing this ratio. PGA and PLGA are hydrolytically degraded through hydrolysis of the ester bonds, but the degradation rate of PLGA is high, and so PLA-PGA copolymer overcomes this limitation of PGA.

Poly(ε-Caprolacton) (PCL)

Poly(ε-caprolacton) (PCL) is a semicrystalline polyester synthesized also by ring-opening polymerization of ε-caprolactone monomer. This polymer is characterized by low cost, its ability to be miscible with a wide range of other polymer solutions, solubility in numerous organic solvents, and it can be melted and molded at low temperatures (55–60 °C). But, very slow degradation rate (2–3 years) is a main limitation of this polymer. This limitation can be solved by copolymerized ε-caprolactone with L-lactide, glycolide, or polyethylene glycol (PEG). Overall, PCL has been approved by FDA for tissue engineering, drug delivery, and other biomedical applications. Compared with previously mentioned PLA and PLGA polymers, PCL is cheaper, has better processability and it has high thermal stability which enables it to be shaped by the melting process.

Polyvinyl Alcohol (PVA)

Polyvinyl alcohol (PVA) is a semi-crystalline linear synthetic polymer. It is synthesized by alkaline hydrolysis (in aqueous sodium hydroxide) of vinyl acetate by replacement of an ester group in vinyl acetate with a hydroxyl group to form a precipitate of PVA. The properties of PVA are determined by the degree of hydrolysis [76]. PVA is widely utilized in the synthesis of vinylon fibers and thermoplastic poly(vinyl butyral) (PVB) which is used as a strong adhesive product. Because of low cost, biocompatibility, controlled biodegradation, hydrophilicity, and crosslinking capability (using monoaldehydes, such as glutaraldehyde, formaldehyde, and acetaldehyde crosslinker), PVA has been used in tissue regeneration, drug delivery and drug encapsulation.

Polyethylene Glycol (PEG)

Polyethylene glycol is a linear polyether polymer synthesized from the reaction of ethylene oxide with water, ethylene glycol, or ethylene glycol oligomers catalyzed with acid or base. Polymer molecular weight is controlled by the ratio of reactants. PEG represents good biocompatibility, nontoxicity, low immunogenicity, and high hydrophilicity, its degradation rate can be tailored via its molecular weight controlling, and solubility in numerous organic solvents and water [77]. Certain molecules (e.g. peptides, proteins, and drugs) have been modified through binding PEG with their end-groups by covalent and non-covalent bonds, this process is called PEGylation [78]. PEGylation is applied to improve drug pharmacokinetic behavior and decrease drug solubility and immunogenicity.

Polyurethane (PU)

Polyurethane (PU) can be exhibited in both thermoplastic and thermoset forms. It is synthesized by a polyaddition reaction between polyol compound (contains ≥ 2 (–OH) groups) and a compound containing ≥ 2 isocyanate groups (–N=C=O). The main characteristic feature of PU is its superior mechanical properties, as well as, it has high biocompatibility, and mechanical, thermal, degradation, or chemical properties can be tailored by varying starting materials and conditions of reaction. PUs have been used in medical applications for a long time (since 1995), and they passed through development, modification, and improvement [79]. Among them, biodegradable PU has been tested in tissue engineering and regenerative medicine, specifically, their mechanical strength and biodegradation can be tailored. However, the main disadvantage of PU is complex and non-ecofriendly synthesis process. Moreover, it was used as denture base resins,

Poly(Methyl Methacrylate) (PMMA)

Poly(methyl methacrylate) (PMMA) is a synthetic thermoplastic polymer synthesized by polymerization of methyl methacrylate monomer using free radical and anionic initiator by bulk, suspension, solution, and emulsion techniques [80]. PMMA is a non-biodegradable polymer, but it possesses good biocompatibility. Accordingly, it has been used for orthopedic bone cements loaded with drugs. Such cement is characterized by high mechanical properties, low cost, biocompatibility, non-toxicity, and minimal inflammation. However, the main drawback of PMMA cements is the high exothermic temperature of polymerization process which can cause thermal necrosis. Furthermore, PMMA has been utilized for bone tissue regeneration as a scaffolding material for load-bearing applications thanks to its high mechanical strength and stability. The nanofibrous scaffold was successfully prepared from PMMA which showed good cell viability [81]. Strategies used to enhance PMMA scaffold biodegradability and bioactivity are, introducing another biodegradable polymer, incorporation of bioactive ceramics (e.g. HAp), and post-treatment of scaffold surface by coating it with biodegradable and bioactive composite [82].

Polyether Ether Ketone (PEEK)

Polyether ether ketone (PEEK) is a thermoplastic semi-crystalline linear, aromatic polymer. It is synthesized by step-growth polymerization of disodium salt of hydroquinone with 4,4'-difluorobenzophenone in aprotic solvents (e.g. diphenyl sulfone) at about 300 °C with an assist of sodium carbonate [83]. It is characterized by superior mechanical strength properties (elastic modulus range 3–4 GPa), high chemical durability, and good biocompatibility. In this respect, it is extremely used in dental applications, such as fixed, removable, metal-free fixed dental, and maxillofacial prostheses. In addition, PEEK has been used in orthopedic, trauma, spine, and surgeries as implants. Also, it has been used in bone tissue engineering, specifically in load-bearing parts. However, low biodegradation and bioinertness limit PEEK application in tissue regeneration. Therefore, there are strategies currently applied to improve PEEK bioactivity, such as surface treatment of the polymer by physical treatment, such as plasma which increases the hydrophilicity of the polymer surface, Chemical treatment, in which the PEEK surface is functionalized by functional groups, such as hydroxyl, carboxyl, and amino groups. For example, a nanostructured 3D porous surface is produced by sulfonation of PEEK and subsequently immersed in water. Surface coating by HAp, TiO_2, nBG, etc. [84].

2.2.4 Composites

Composite biomaterials are materials that are made from two or more types of different materials. Materials forming the composite are separated in a nanoscale,

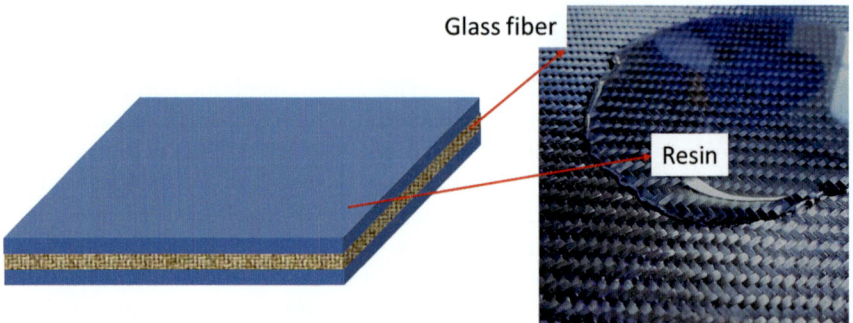

Fig. 2.12 Example of composite material based on glass fiber textile and resin

microscale, or macroscale scale. Composite combines the properties of constituent materials to give material with new, essentially, enhanced properties (mechanical, chemical resistance, biocompatibility, etc.), such as fiber-reinforced epoxy resin (Fig. 2.12).

The composite biomaterials have been released and developed to obtain new materials with improved properties and combine the advantages of each composite-made phase. An example of composite biomaterials is a combination of bioactive ceramics (e.g. HAp and bioglass) and bioineret polymers (e.g. PMMA polymer) to improve the osteoconductivity, bioactivity, and biocompatibility of the polymer, and enhance ceramic mechanical properties on the other side. Accordingly, the development of new composite biomaterials as a third generation of biomaterials, specifically for tissue regeneration, is progressively increasing every year. The majority of composite biomaterials are based on the polymer phase and bioactive ceramic phase. In the following paragraphs, we'll demonstrate some of the previous studies concerned with composite biomaterials for tissue regeneration.

Calcium phosphate composite scaffolds used for bone regeneration have been intensively studied. Venugopal et al., studied the mineralization of osteoblasts on HAp/collagen nanofibrous scaffolds by electrospinning. The results showed that the HAp-contained scaffold more enhanced the mineralization of osteoblasts than HAp-free scaffold [85]. Sotome et al., utilized hydroxyapatite/type 1 collagen composite scaffold for bone regeneration. The scaffold was characterized by good elasticity with a degree to it can be handled and cut with a knife (Fig. 2.13). The authors tested and implanted their materials in 63 patients with bone defects caused by fractures, benign bone tumors, or harvesting of autografts and followed up the patients during 24 weeks after surgery. In contrast, they utilized porous β-TCP for comparison. HAp/collagen higher bone regeneration grade than that of porous β-TCP. But, the incidence of adverse effects resulting from HAp/collagen was higher than that caused by β-TCP [86]. Yu et al., synthesized nano-HAp/silk fibroin scaffolds with naringin (it is a natural flavonoid glycoside widely used in osteoarthritis and osteoporosis treatment). They studied the bone healing ability of such materials and bone defects in rats. Their findings presented that the scaffolds demonstrated good biocompatibility and cell viability, and naringin-contained scaffold enhanced in vitro osteogenic differentiation. As well, the bone defects completely closed in 16 weeks [87]. Harb et al., fabricated β-TCP/PCL/nano-ceria composite scaffolds by rapid prototyping technique for bone regeneration. Incorporation of nano-ceria with 10 wt% enhanced protein adsorption, increased degradation, and extended cell proliferation of the scaffolds, and it favored the osteogenic differentiation of mesenchymal stem cells. [88]. Huang et al., prepared anticancer and antibacterial selenium/strontium/zinc-doped HAp/PCL scaffolds. The scaffolds showed antitumor (due to presence of selenium), osteogenic (due to presence of strontium and zinc), and antibacterial (due to presence of zinc) capacities. In addition, in vivo tests presented that the scaffolds enhanced bone healing of rat femoral defects after 3 months of implantation, and they significantly suppress tumors.

Bioactive glass (BG), as mentioned before, is an osteoinductive material and it can form a bond with the host tissue and promote osteogenesis and angiogenesis. Because of its distinctive biological properties, BG has been combined with biodegradable and non-biodegradable polymers to improve their bioactivity. El-Kady et al., studied the effect of changing the percentage of nBG on different properties of nBG/PLLA composite scaffold. The results showed that nBG decreased porosity and pore size, but increased degradation rate and the water uptake. In addition, nBG enhanced in vitro bioactivity of the scaffolds [89]. In an extended work, they synthesized 45S5 glass/PLLA composite scaffolds by freeze-extraction technique and studied the effect of glass content on biodegradation and bioactivity. Likewise, apparent porosity and pore size decreased as glass content increased, whereas, glass particles induced the formation of HAp crystals on the scaffold surfaces [90]. de Souza et al., prepared BG/poly(L-lactide-co-ε-caprolactone)/polyethylene glycol composite scaffold by electrospinning. The results showed that polyethylene glycol and BG

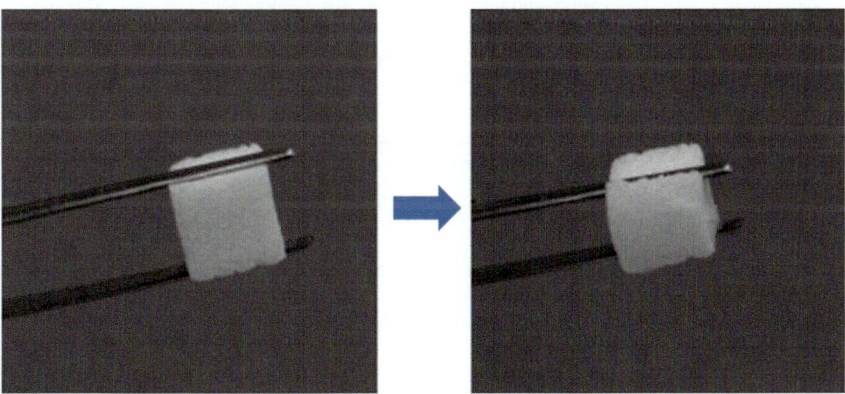

Fig. 2.13 Elasticity of wet porous HAp/collagen scaffold shows good handling during surgery. Modified from [86]. Copyright Elsevier, 2016. This article is an open access licensed under a Creative Commons Attribution 4.0 International License

increased scaffold water absorption. Moreover, BG improved the mechanical properties, thermal stability, cell adhesion, and osteoinductive potential of the scaffold. Hammoda et al., examined the effect of Ce-doped nBG/collagen/chitosan scaffolds on cell morphology and proliferation of rabbit's bone marrow mesenchymal stem (BM-MSCs). The results demonstrated that released cerium ions stimulated early osteogenic proliferation and hence bone mineralization [91]. In a complementary study, the authors studied the anticancer activity of the same scaffolds. The scaffolds containing nBG doped with 5 mol.% cerium showed the highest cytotoxic effect against osteosarcoma cells [92]. Janmohammadi et al., evaluated the ability of BG/PCL/tragacanth gum scaffold to form new bone in rat calvarial defect. Tragacanth (TG) is an anionic polysaccharide natural gum that can increase cell adhesion, spread, and differentiation. The histological analysis showed that BG/PCL/TG scaffold enhanced bone formation indexed by the promotion of osteogenesis-related gene expression (e.g. Runx2 and collagen type I). Hence, BG and TG enhanced new bone formation and mineralization, and they increased scaffold hydrophilicity also [93]. Manjubaashini et al., prepared Na-free 45S5 BG/chitosan scaffolds by lyophilization technique for diabetic wound healing. They implanted the scaffolds in diabetic Sprague Dawley rats. The results presented that the scaffold contained BG showed enhanced wound healing, and the wound completely closed after only 14 days.

References

1. Anneroth G, Ericsson A, Zetterqvist L (1990) Tissue integration of Al_2O_3-ceramic dental implants (Frialit)-a case report. Swed Dent J 14:63–70
2. Yanyan S, Guangxin W, Guoqing S, Yaming W, Wuhui L, Osaka A (2020) Effects of amino acids on conversion of calcium carbonate to hydroxyapatite. RSC Adv 10:37005–37013
3. Singh R, Bathaei MJ, Istif E, Beker L (2020) A review of bioresorbable implantable medical devices: materials, fabrication, and implementation. Adv Healthc Mater 9:2000790
4. Callister WD, Rethwisch DG, Blicblau A, Bruggeman K, Cortie M, Long J et al (2007) Materials science and engineering: an introduction. Wiley, New York
5. Blum IR, Jagger DC, Wilson NH (2011) Defective dental restorations: to repair or not to repair? Part 2: All-ceramics and porcelain fused to metal systems. Dent Update 38:150–158
6. Maji A, Choubey G (2018) Microstructure and mechanical properties of alumina toughened zirconia (ATZ). Mater Today Proc 5:7457–7465
7. Driessens F (2018) Formation and stability of calcium phosphates in relation to the phase composition of the mineral in calcified tissues. In: Bioceramics of calcium phosphate. CRC Press, pp 1–32
8. Dorozhkin SV (2019) Functionalized calcium orthophosphates ($CaPO_4$) and their biomedical applications. J Mater Chem B 7:7471–7489
9. Bohner M (2010) Design of ceramic-based cements and putties for bone graft substitution. Eur Cell Mater 20:1–12
10. Dorozhkin SV (2016) Multiphasic calcium orthophosphate ($CaPO_4$) bioceramics and their biomedical applications. Ceram Int 42:6529–6554
11. Anthony JW, Bideaux RA, Bladh KW, Nichols MC (2001) Handbook of mineralogy. Mineral Data Publishing
12. Arcos D, Rodríguez-Carvajal J, Vallet-Regí M (2004) The effect of the silicon incorporation on the hydroxylapatite structure. A neutron diffraction study. Solid State Sci 6:987–994

13. Fathi MH, Zahrani EM (2009) Fabrication and characterization of fluoridated hydroxyapatite nanopowders via mechanical alloying. J Alloy Compd 475:408–414
14. Abdelraof M, Farag MM, Al-Rashidy ZM, Ahmed HY, El-Saied H, Hasanin MS (2022) Green synthesis of bioactive hydroxyapatite/cellulose composites from food industrial wastes. J Inorg Organomet Polym Mater 32:4614–4626
15. Catauro M, Barrino F, Blanco I, Piccolella S, Pacifico S (2020) Use of the sol–gel method for the preparation of coatings of titanium substrates with hydroxyapatite for biomedical application. Coatings 10:203
16. Zhang G, Chen J, Yang S, Yu Q, Wang Z, Zhang Q (2011) Preparation of amino-acid-regulated hydroxyapatite particles by hydrothermal method. Mater Lett 65:572–574
17. Zhang H-b, Zhou K-c, Li Z-y, Huang S-p, Zhao Y-z (2009) Morphologies of hydroxyapatite nanoparticles adjusted by organic additives in hydrothermal synthesis. J Cent South Univ Technol 16:871–875
18. Aizawa M, Hanazawa T, Itatani K, Howell F, Kishioka A (1999) Characterization of hydroxyapatite powders prepared by ultrasonic spray-pyrolysis technique. J Mater Sci 34:2865–2873
19. Mondal S, Hoang G, Manivasagan P, Moorthy MS, Kim HH, Vy Phan TT et al (2019) Comparative characterization of biogenic and chemical synthesized hydroxyapatite biomaterials for potential biomedical application. Mater Chem Phys 228:344–356
20. Bohner M (2007) Reactivity of calcium phosphate cements. J Mater Chem 17:3980–3986
21. Maguire ME, Cowan JA (2002) Magnesium chemistry and biochemistry. Biometals 15:203–210
22. Ostrowski N, Roy A, Kumta PN (2016) Magnesium phosphate cement systems for hard tissue applications: a review. ACS Biomater Sci Eng 2:1067–1083
23. Moseke C, Saratsis V, Gbureck U (2011) Injectability and mechanical properties of magnesium phosphate cements. J Mater Sci Mater Med 22:2591–2598
24. Walling SA, Provis JL (2016) Magnesia-based cements: a journey of 150 years, and cements for the future? Chem Rev 116:4170–4204
25. Wagh AS, Primus C (2006) Method and product for phosphosilicate slurry for use in dentistry and related bone cements. Google Patents
26. Farag MM, Yun H (2014) Effect of gelatin addition on fabrication of magnesium phosphate-based scaffolds prepared by additive manufacturing system. Mater Lett 132:111–111
27. Jongman Lee MMF, Park EK, Lim J, Yun H (2014) A simultaneous process of 3D magnesium phosphate scaffold fabrication and bioactive substance loading for hard tissue regeneration. Mater Sci Eng C 36:52–260
28. Tamimi F, Le Nihouannen D, Bassett DC, Ibasco S, Gbureck U, Knowles J et al (2011) Biocompatibility of magnesium phosphate minerals and their stability under physiological conditions. Acta Biomater 7:2678–2685
29. Yu Y, Wang J, Liu C, Zhang B, Chen H, Guo H et al (2010) Evaluation of inherent toxicology and biocompatibility of magnesium phosphate bone cement. Colloids Surf B 76:496–504
30. Mestres G, Abdolhosseini M, Bowles W, Huang S-H, Aparicio C, Gorr S-U et al (2013) Antimicrobial properties and dentin bonding strength of magnesium phosphate cements. Acta Biomater 9:8384–8393
31. Mestres G, Ginebra MP (2011) Novel magnesium phosphate cements with high early strength and antibacterial properties. Acta Biomater 7:1853–1861
32. Soudée E, Péra J (2000) Mechanism of setting reaction in magnesia-phosphate cements. Cem Concr Res 30:315–321
33. Lee J, Farag MM, Park EK, Lim J, Yun HS (2014) A simultaneous process of 3D magnesium phosphate scaffold fabrication and bioactive substance loading for hard tissue regeneration. Mater Sci Eng C Mater Biol Appl 36:252–260
34. Farag MM, Yun H (2014) Effect of gelatin addition on fabrication of magnesium phosphate-based scaffolds prepared by additive manufacturing system. Mater Lett 132:111–115
35. Ginebra M, Boltong M, Driessens F, Bermudez O, Fernández E, Planell J (1994) Preparation and properties of some magnesium-containing calcium phosphate cements. J Mater Sci Mater Med 5:103–107

36. Nabiyouni M, Brückner T, Zhou H, Gbureck U, Bhaduri SB (2018) Magnesium-based bioceramics in orthopedic applications. Acta Biomater 66:23–43
37. Shearer A, Montazerian M, Sly JJ, Hill RG, Mauro JC (2023) Trends and perspectives on the commercialization of bioactive glasses. Acta Biomater 160:14–31
38. ASTM (1945) Standard terminology of glass and glass products. American Society for Testing and Materials, WC
39. Montazerian M, Dutra ZE (2016) History and trends of bioactive glass-ceramics. J Biomed Mater Res Part A 104:1231–1249
40. Hench LL, Splinter RJ, Allen W, Greenlee T (1971) Bonding mechanisms at the interface of ceramic prosthetic materials. J Biomed Mater Res 5:117–141
41. Hench LL (1998) Bioactive materials: the potential for tissue regeneration. J Biomed Mater Res 41:511–518
42. Rahaman MN, Day DE, Bal BS, Fu Q, Jung SB, Bonewald LF et al (2011) Bioactive glass in tissue engineering. Acta Biomater 7:2355–2373
43. Brézulier D, Chaigneau L, Jeanne S, Lebullenger R (2021) The challenge of 3D bioprinting of composite natural polymers PLA/bioglass: trends and benefits in cleft palate surgery. Biomedicines 9:1553
44. Hench L, Paschall H (1974) Histochemical responses at a biomaterial's interface. J Biomed Mater Res 8:49–64
45. Hench LL, Andersson O (1993) Bioactive glasses. Adv Ser Ceram 1:41–62
46. Kokubo T, Ito S, Huang ZT, Hayashi T, Sakka S, Kitsugi T et al (1990) Ca, P-rich layer formed on high-strength bioactive glass-ceramic A-W. J Biomed Mater Res 24:331–343
47. Andersson Ö, Liu G, Kangasniemi K, Juhanoja J (1992) Evaluation of the acceptance of glass in bone. J Mater Sci Mater Med 3:145–150
48. Farag MM (2023) Recent trends on biomaterials for tissue regeneration applications: review. J Mater Sci 58:527–558
49. Marion NW, Liang W, Liang W, Reilly GC, Day DE, Rahaman MN et al (2005) Borate glass supports the in vitro osteogenic differentiation of human mesenchymal stem cells. Mech Adv Mater Struct 12:239–246
50. Yao A, Wang D, Huang W, Fu Q, Rahaman MN, Day DE (2007) In vitro bioactive characteristics of borate-based glasses with controllable degradation behavior. J Am Ceram Soc 90:303–306
51. Fu Q, Rahaman MN, Fu H, Liu X (2010) Silicate, borosilicate, and borate bioactive glass scaffolds with controllable degradation rate for bone tissue engineering applications. I. Preparation and in vitro degradation. J Biomed Mater Res Part A 95:71–164
52. Huang W, Day DE, Kittiratanapiboon K, Rahaman MN (2006) Kinetics and mechanisms of the conversion of silicate (45S5), borate, and borosilicate glasses to hydroxyapatite in dilute phosphate solutions. J Mater Sci Mater Med 17:583–596
53. Omar AE, Ibrahim AM, Abd El-Aziz TH, Al-Rashidy ZM, Farag MM (2021) Role of alkali metal oxide type on the degradation and in vivo biocompatibility of soda-lime-borate bioactive glass. J Biomed Mater Res B Appl Biomater 109:1059–1073
54. El-Meliegy E, Farag MM, Knowles JC (2016) Dissolution and drug release profiles of phosphate glasses doped with high valency oxides. J Mater Sci Mater Med 27:108
55. Vallet-Regí M, Ragel C, Salinas AJ (2003) Glasses with medical applications. Eur J Inorg Chem 2003:1029–1042
56. Li R, Clark A, Hench L (1991) An investigation of bioactive glass powders by sol-gel processing. J Appl Biomater 2:231–239
57. De Aza P, Guitian F, Merlos A, Lora-Tamayo E, De Aza S (1996) Bioceramics—simulated body fluid interfaces: pH and its influence of hydroxyapatite formation. J Mater Sci Mater Med 7:399–402
58. Brinker CJ, Scherer GW (2013) Sol-gel science: the physics and chemistry of sol-gel processing. Academic Press
59. McDonagh C, Sheridan F, Butler T, MacCraith B (1996) Characterisation of sol-gel-derived silica films. J Non-Cryst Solids 194:72–77

60. Schatkoski VM, Larissa do Amaral Montanheiro T, Canuto de Menezes BR, Pereira RM, Rodrigues KF, Ribas RG et al (2021) Current advances concerning the most cited metal ions doped bioceramics and silicate-based bioactive glasses for bone tissue engineering. Ceram Int 47:2999–3012

61. Awais M, Aizaz A, Nazneen A, Bhatti QuA, Akhtar M, Wadood A et al (2022) A review on the recent advancements on therapeutic effects of ions in the physiological environments. Prosthesis 4:263–316

62. Fratzl P, Misof K, Zizak I, Rapp G, Amenitsch H, Bernstorff S (1998) Fibrillar structure and mechanical properties of collagen. J Struct Biol 122:119–122

63. Matinong AME, Chisti Y, Pickering KL, Haverkamp RG (2022) Collagen extraction from animal skin. Biology 11:905

64. Lou X, Chirila TV (1999) Swelling behavior and mechanical properties of chemically cross-linked gelatin gels for biomedical use. J Biomater Appl 14:184–191

65. Prasertsung I, Mongkolnavin R, Damrongsakkul S, Wong C (2010) Surface modification of dehydrothermal crosslinked gelatin film using a 50 Hz oxygen glow discharge. Surf Coat Technol 205:S133–S138

66. Weisel JW, Litvinov RI (2013) Mechanisms of fibrin polymerization and clinical implications. Blood J Am Soc Hematol 121:1712–1719

67. Bielecki T, Dohan Ehrenfest DM (2012) Platelet-rich plasma (PRP) and platelet-rich fibrin (PRF): surgical adjuvants, preparations for in situ regenerative medicine and tools for tissue engineering. Current Pharm Biotechnol 13:1121–1130

68. Yatmaz E, Turhan I (2015) Hyaluronic acid and production by fermentation

69. Shi C, Zhu Y, Ran X, Wang M, Su Y, Cheng T (2006) Therapeutic potential of chitosan and its derivatives in regenerative medicine. J Surg Res 133:185–192

70. Rinaudo M (2006) Chitin and chitosan: properties and applications. Prog Polym Sci 31:603–632

71. Martinac A, Filipović-Grčić J, Voinovich D, Perissutti B, Franceschinis E (2005) Development and bioadhesive properties of chitosan-ethylcellulose microspheres for nasal delivery. Int J Pharm 291:69–77

72. Uyen NTT, Hamid ZAA, Tram NXT, Ahmad N (2020) Fabrication of alginate microspheres for drug delivery: a review. Int J Biol Macromol 153:1035–1046

73. Kumar B, Singh N, Kumar P (2024) A review on sources, modification techniques, properties and potential applications of alginate-based modified polymers. Eur Polym J 113078

74. Bhatia S, Bhatia S (2016) Natural polymers vs synthetic polymer. In: Natural polymer drug delivery systems: nanoparticles, plants, and algae, pp 95–118

75. Middleton JC, Tipton AJ (2000) Synthetic biodegradable polymers as orthopedic devices. Biomaterials 21:2335–2346

76. Halima NB (2016) Poly (vinyl alcohol): review of its promising applications and insights into biodegradation. RSC Adv 6:39823–39832

77. Kong X, Tang Q, Chen X, Tu Y, Sun S, Sun Z-l (2017) Polyethylene glycol as a promising synthetic material for repair of spinal cord injury. Neural Regener Res 12:1003–1008

78. Turecek PL, Bossard MJ, Schoetens F, Ivens IA (2016) PEGylation of biopharmaceuticals: a review of chemistry and nonclinical safety information of approved drugs. J Pharm Sci 105:460–475

79. Bos M, Van Dam GW, Jongsma T, Bruin P, Pennings AJ (1995) The effect of filler surface modification on the mechanical properties of hydroxyapatite-reinforced polyurethane composites. Compos Interfaces 3:169–176

80. Yuan M, Huang D, Zhao Y (2022) Development of synthesis and application of high molecular weight poly (methyl methacrylate). Polymers 14:2632

81. Ura DP, Karbowniczek JE, Szewczyk PK, Metwally S, Kopyściański M, Stachewicz U (2019) Cell integration with electrospun PMMA nanofibers, microfibers, ribbons, and films: a microscopy study. Bioengineering 6:41

82. Rahmani-Monfard K, Fathi A, Rabiee SM (2016) Three-dimensional laser drilling of poly-methyl methacrylate (PMMA) scaffold used for bone regeneration. Int J Adv Manuf Technol 84:2649–2657

83. Kemmish D (2010) Update on the technology and applications of polyaryletherketones. iSmithers, Smithers Rapra Publishing, Shropshire

84. Ma R, Tang T (2014) Current strategies to improve the bioactivity of PEEK. Int J Mol Sci 15:5426–5445

85. Venugopal J, Low S, Choon AT, Sampath Kumar TS, Ramakrishna S (2008) Mineralization of osteoblasts with electrospun collagen/hydroxyapatite nanofibers. J Mater Sci Mater Med 19:2039–2046

86. Sotome S, Ae K, Okawa A, Ishizuki M, Morioka H, Matsumoto S et al (2016) Efficacy and safety of porous hydroxyapatite/type 1 collagen composite implantation for bone regeneration: a randomized controlled study. J Orthop Sci 21:373–380

87. Yu X, Shen G, Shang Q, Zhang Z, Zhao W, Zhang P et al (2021) A naringin-loaded gelatin-microsphere/nano-hydroxyapatite/silk fibroin composite scaffold promoted healing of critical-size vertebral defects in ovariectomised rat. Int J Biol Macromol

88. Harb SV, Kolanthai E, Pugazhendhi AS, Beatrice CAG, Pinto LA, Neal CJ et al (2024) 3D printed bioabsorbable composite scaffolds of poly (lactic acid)-tricalcium phosphate-ceria with osteogenic property for bone regeneration. Biomater Biosyst 13:100086

89. El-Kady AM, Ali AF, Farag MM (2010) Development, characterization, and in vitro bioactivity studies of sol–gel bioactive glass/poly(l-lactide) nanocomposite scaffolds. Mater Sci Eng C 30:120–131

90. El-Kady AM, Saad EA, El-Hady BMA, Farag MM (2010) Synthesis of silicate glass/poly(l-lactide) composite scaffolds by freeze-extraction technique: characterization and in vitro bioactivity evaluation. Ceram Int 36:995–1009

91. Hammouda HF, Farag MM, El Deftar MM, Abdel-Gabbar M, Mohamed BM (2022) Effect of Ce-doped bioactive glass/collagen/chitosan nanocomposite scaffolds on the cell morphology and proliferation of rabbit's bone marrow mesenchymal stem cells-derived osteogenic cells. J Genet Eng Biotechnol 20:33

92. Hammouda H, Farag M, El Deftar M, Abdel-Gabbar M, Mohamed B (2022) Ce-doped nanobioactive glass/collagen/chitosan composite scaffolds: biocompatibility with normal rabbit's osteoblast cells and anticancer activity test. Adv Anim Vet Sci 10:712–724

93. Janmohammadi M, Doostmohammadi N, Bahraminasab M, Nourbakhsh MS, Arab S, Asgharzade S et al (2024) Evaluation of new bone formation in critical-sized rat calvarial defect using 3D printed polycaprolactone/tragacanth gum-bioactive glass composite scaffolds. Int J Biol Macromol 270:132361

Chapter 3
Methods Used to Fabricate Tissue Engineering Scaffolds

Abstract This chapter represents different methods used for scaffold fabrication. Scaffolds can be fabricated by several methods. The variety of scaffold fabrication methods came from the development of the tissue engineering field which forced the scientists to search for advanced methods to prepare ideal scaffold. Meanwhile, the required properties of the scaffold applied for specific applications determine the fabrication method. In this chapter, we illustrate the most important methods for scaffold fabrication accompanied by simple practical examples from our practical experience or previous literature. The fabrication methods of the scaffold can be generally subdivided into two techniques; conventional techniques and rapid prototyping (RP) techniques (Fig. 3.1).

3.1 Conventional Fabrication Techniques

3.1.1 Freeze-Drying

Freeze-drying basic depends mainly on the sublimation concept. In this technique, the polymer is dissolved in a solvent (aqueous or nonaqueous), and then the mixture is frozen under the freezing point of the polymer solution, at this point the solvent becomes in a solid state. The frozen sample is put into to freeze-dryer device in which the solvent in the solid state transforms to a gaseous state by sublimation living behind polymer with interconnected pores forming a porous scaffold [1]. Freeze-dryer is composed of four main parts; vacuum drying chamber, heat source, condenser with refrigeration system, and vacuum pump (Fig. 3.2). The advantages of this technique are fast scaffold preparation, the scaffold is characterized by a highly porous structure, the pore shape and size can be controlled by controlling the cooling protocol, and low-temperature fabrication of the scaffold which is suitable for biological compounds. However, the pore shape in the scaffold produced in this technique is irregular pore sizes, it consumes high energy, and usually, the polymer is dissolved in toxic solvents that are cytotoxic. Table 3.1 shows the advantages and disadvantages of different conventional techniques (Fig. 3.1).

Table 3.1 Advantages and disadvantages of different conventional scaffold fabricating techniques

Technique	Advantages	Disadvantages
Freeze-drying	Fast scaffold preparation Highly porous scaffold structure. Controlled pore shape and size Low-temperature fabrication	Irregular pore sizes Consumption of high energy Toxic solvents are usually used
Solvent casting and particle leaching	High porosity (up to 90%) Tailored pore size Simple and cost-effective method	Time consuming Traces of porogen can be remained The scaffold thickness is limited Formation of skin on the scaffold surface
Gas foaming	Simple and cheap method Porosity range 85–93% Pore size range 30–700 μm The scaffold is free from the skin layer	Not suitable for complex shapes Pore size cannot be controlled
Electrospinning	Simple and fast Produces nanofibrous scaffolds more similar to ECM	Difficult to prepare complex shapes Inhomogeneous pore size and shape using toxic solvents
Thermal-induced phase separation	3D nanofibrous scaffolds can be prepared Simple and cost-effective method	Limited types of polymer Using of toxic solvent Time consuming

The following chitosan/collagen/nBG scaffolds fabrication steps are shown as an example to represent the procedure as shown in Fig. 3.3. Firstly, collagen and chitosan solutions were mixed with 1:1 volume ratio, and then 30 wt% of nBG particles were added to this polymer mixture stirred for 3 h, and ultrasonicated for 30 min. The polymer/glass mixture was frozen at $-20\ °C$ and put in a freeze-dryer adjusted at $-50\ °C$ for 2 days to attain the final scaffold [2].

3.1.2   Solvent Casting and Particle Leaching

In this technique, salt particles (porogen) with definite particle sizes are distributed homogeneously in the polymer dissolved in a solvent solution (aqueous or organic solvent). Then, the mixture is cast in the mold and left in the air or dryer to evaporate the solvent forming salt particles uniformly distributed in the polymer matrix. The salt particles are leached by putting the polymer/salt in a solvent (usually water) suitable to dissolve salt thereafter. Alternatively, the solvent used to dissolve the polymer can be avoided by casting the polymer melt on porogen and subjected to pressing. Then

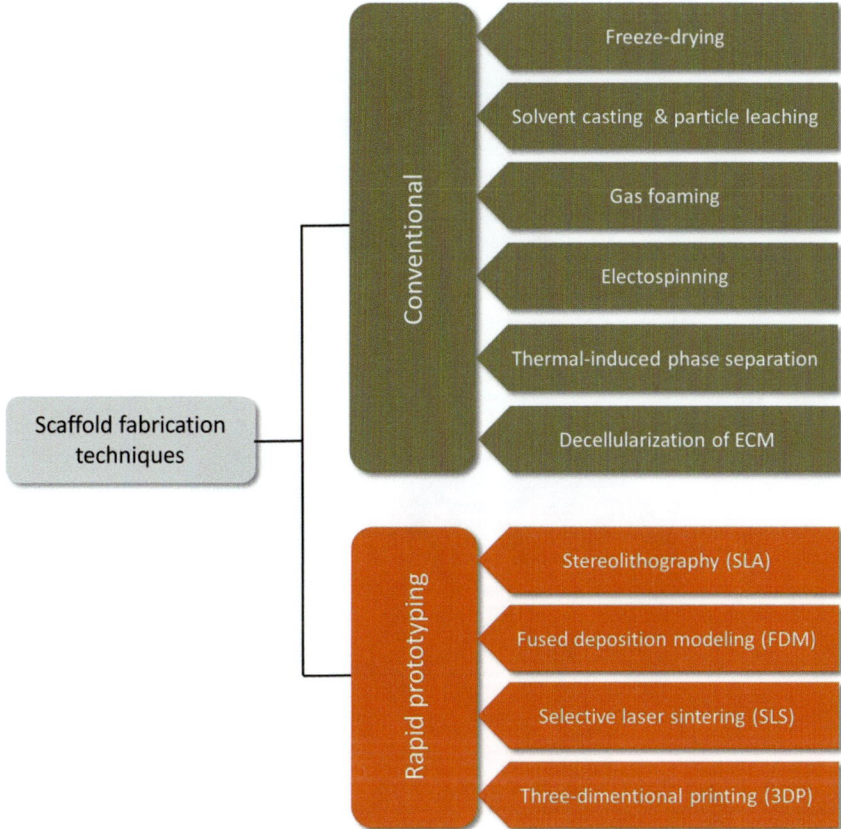

Fig. 3.1 Different techniques are used for scaffold fabrication. They are subdivided into two major groups; conventional techniques and rapid prototyping (RP) techniques

the porogen is leached out following the above procedure. The advantages of solvent casting and salt (porogen) leaching are; obtaining a scaffold with high porosity (up to 90%), pore size can be tailored by changing salt particle size, and it is a simple and cost-effective method. the disadvantages of this method are that; consumes of long time to reach complete porogen leaching, traces of porogen can remain in the scaffold which affects the scaffold cytocompatibility, the scaffold thickness is limited, skin can be formed on the scaffold surface which hinders the cells to penetrate the scaffold, low degree of interconnectivity [3].

Thadavirul et al., prepared PCL scaffold by solvent casting and porogen leaching. They used two porogens; NaCl and PEG polymer (Fig. 3.4). Firstly, they added PCL and PEG with 1:1 weight ratio to chloroform. After the complete dissolution of PCL, NaCl with a particle size range 400–500 μm was added and mixed well. The PCL/ porogens cast in mould and left to dry. After complete drying, it was immersed in distilled water for 2 days with periodical refresh the water every 8 h, and finally, the formed scaffold was dried [4].

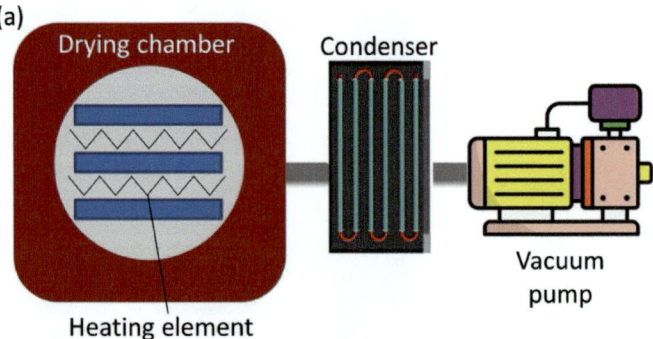

Fig. 3.2 **a** Schematic diagram of freeze-dryer components, and **b** real freeze-dryer

3.1.3 Gas Foaming

This method is based on a generation of inert gases, such as carbon dioxide and ammonia in a semi-solidified polymer. Where, salt particles, such as ammonium bicarbonate, are dispersed homogeneously in polymer–solvent and shaped into the desired shape. When the polymer reaches a gel-like state, it is immersed in an acidic solution or hot water to induce the rising of carbon dioxide and ammonia gases which cause the bubbling of the polymer producing highly interconnected porous architecture [5]. Meanwhile, as ammonium bicarbonate particles dissolve during this process they leave behind pores, and so this method can be considered a combining of gas foaming and salt leaching methods. The porosity and pore size range of the scaffold prepared by this method can be 85–93% and 30–700 μm, respectively. Unlike the salt leaching method, the scaffold is free from the skin layer, thus permitting cell migration from the surface to the scaffold body.

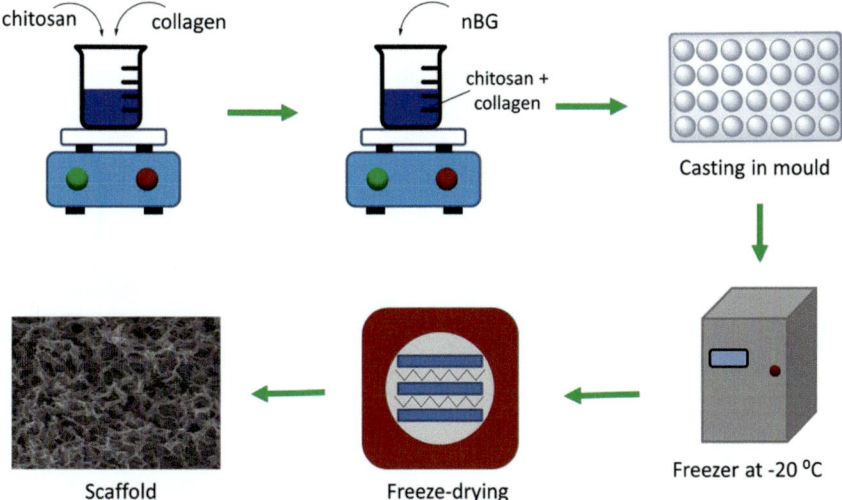

Fig. 3.3 Schematic diagram of preparation of scaffold by freeze-drying technique

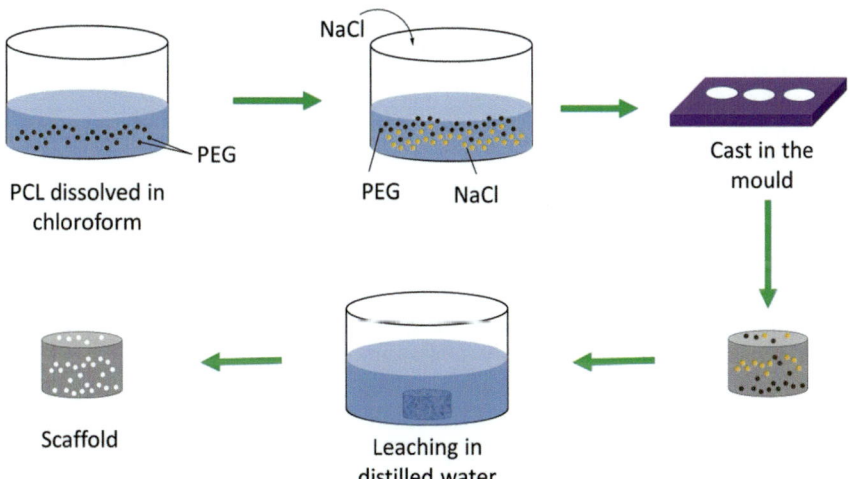

Fig. 3.4 Schematic diagram of preparation of scaffold by solvent casting and porogen leaching technique

The following procedure is shown as an example to represent the process as shown in Fig. 3.5, which depends on our practical experience or previous literature. Herein, we will prepare PCL scaffold using the gas foaming/salt leaching method. Ammonium bicarbonate particles are ground and sieved to get a particle size range of 200–300 μm. 5% (w/v) of PCL of molecular weight 80,000 was prepared by dissolving the polymer in chloroform. After good mixing, some amount of cold ethanol is added

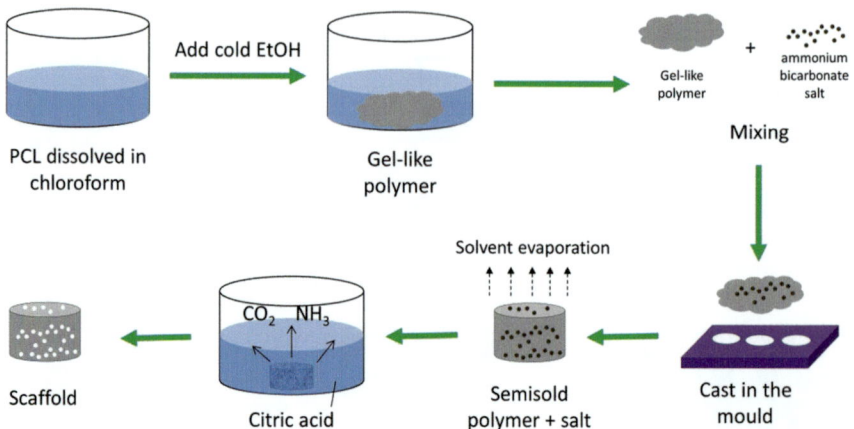

Fig. 3.5 Schematic diagram of preparation of scaffold by gas foaming/salt-leaching method

to the polymer solution to form a gel-like polymer, and then ammonium carbonate particles are mixed with the above polymer gel to obtain a homogeneous paste. The paste is shaped in Teflon mold and let it to dry to become a semi-solidified mixture. Finally, immerse polymer/salt mixture in a supersaturated solution of citric acid until complete effervescence of carbon dioxide and ammonia gases. The final scaffold is washed with distilled water several times and dried.

3.1.4 Electrospinning

The concept of electrospinning technique is an application of high voltage electricity on the liquid droplet to overcome its surface tension and elongates it to nanofibers. The electrospinning device is mainly composed of a high-voltage power supply, a syringe pump to extrude the liquid droplets, and an electroconductive collector which is a static or rotating collector (Fig. 3.6). The high voltage is applied between the syringe needle and collector. Before application of high voltage, the liquid ejects as droplets from the syringe needle tip due to the liquid surface tension. After the application of high voltage, the charge accumulates on the droplet surface causing electrostatic repulsion force to exceed the surface tension force which deforms the droplet's circular shape into a conical shape known as a Taylor cone. After the instant formation of Taylor cone, the liquid is jetted in continuous fibers which are solidified to form a nonwoven mat [6]. The parameters controlling the shape and diameter of electrospun fibers can be divided into two main parameters:

(1) Solution parameters include temperature, type (polar or nonpolar), viscosity, surface tension, and molecular weight.

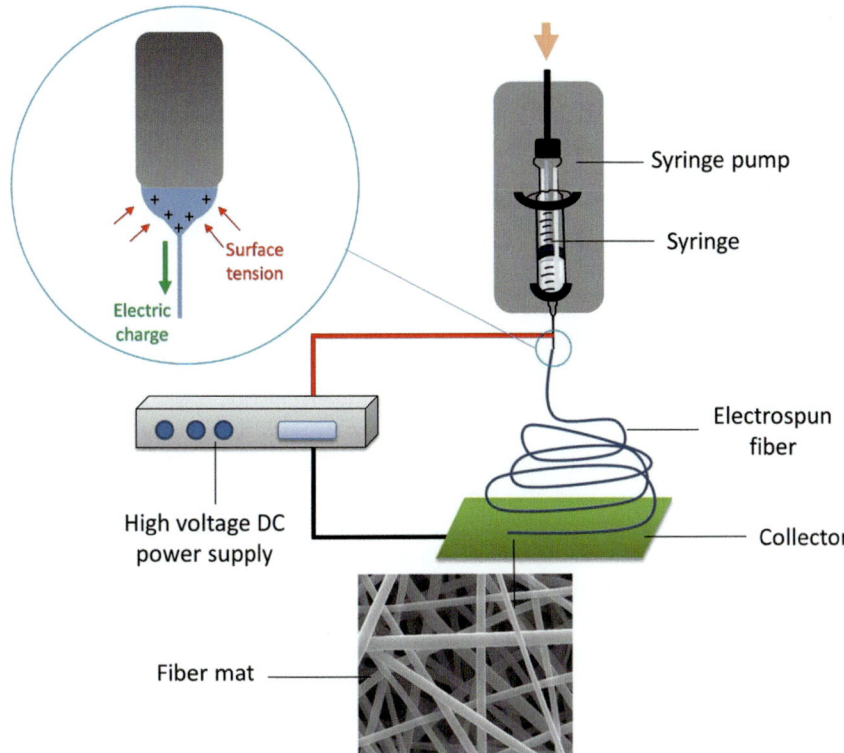

Fig. 3.6 Electrospinning device setup, and the two opposite forces (surface tension and electric charge) applied on the droplet shape at the syringe needle tip

(2) Process parameters include applied voltage, humidity, needle-collector distance, and flow rate.

The main advantages of electrospinning technique are; simplicity, quick, and it produces nanofibrous scaffolds more similar to ECM than scaffolds prepared by the other methods [7]. However, electrospinning technique has some drawbacks such as difficulty in preparing complex shapes, the pore size and shape being inhomogeneous, and complications of cell seeding and migration due to the pore size in a nanoscale [8].

For instance, Gautam et al., prepared gelatin-PCL-nHAp composite scaffold by electrospinning technique. They dissolved 8% w/v gelatin in 80% v/v acetic acid, and dissolved 20% w/v PCL separately in chloroform/methanol solvent (3:1, v/v). Then, they mixed the two solutions by volume ratio 20:80 of gelatin: PCL and kept at room temperature for 3 days. Gelating/PCL solution filled in 3 ml plastic syringe put in syringe pump of electrospinning device. The conditions of electrospinning process were 22 kV, the solution flow rate was 0.1 ml/h, the needle-collector distance was 10 cm, 45% humidity, room temperature, and the collector was an aluminium plate.

After the formation of the polymer nanofibrous scaffold, the scaffold was immersed in 1 wt% of nHAp solution and stirred for 10, 20, and 30 min. finally, the scaffold was washed with distilled water and dried [9].

3.1.5   Thermal-Induced Phase Separation

The base of this technique is inducing phase separation, either liquid–liquid or solid–liquid phase separation. Herein, the polymer is dissolved in a solvent, and then the polymer solution is frozen at a freezing point less than that of the polymer solution. The freezing step causes a separation of polymer solution into two phases; polymer-rich phase and polymer-poor phase (it can be said solvent-rich phase). Solvent-rich phase from solvent crystals when frozen, these crystals extracted by another solvent of lower freezing temperature than that of polymer solution, such as ethanol. Where the frozen polymer solution is immersed in the extracted solution under the same freezing temperature as the polymer solution [6]. This technique is characterized possibility of preparing 3D nanofibrous scaffolds. However, polymers used in fabrication of scaffolds using this technique are limited due to the solvent used to dissolve these polymers. Moreover, the toxic solvent may be used to dissolve the polymer, when traces of solvent in the scaffold likely, cause cytotoxic complications.

El-Kady et al., prepared 45S5 glass/PLA composite scaffold by thermal-induced phase separation. They dissolved 12.5% (w/v) PLA in chloroform, and after complete dissolution of polymer glass particles were added to the polymer solution and well-dispersed by ultrasonic to get a homogenous mixture. Then, the mixture was cast in a cylindrical mold and frozen at -80 °C in the ultra-deep freezer which induced separation of two phases; PLA-rich phase and chloroform crystals-rich phase. The casted frozen glass/polymer samples were soaked in absolute ethanol solution at the same temperature, -80 °C, for 7 days and refreshed 3 times a day for 2 days. After the complete extraction of chloroform, the scaffolds were removed from ethanol and dried at room temperature [10]. Figure 3.7 shows an optical photo and SEM micrograph of one scaffold prepared in this work.

3.1.6   Decellularization of Extracellular Matrix (dECM)

Decellularization method is used to fabricate scaffold mimics tissue microenvironment. In this method, cellular components are removed from tissue or organ to obtain a structural extracellular matrix (ECM) template that maintains the biomimetic ECM microenvironment [11]. Figure 3.8 shows examples of decellularized organ scaffolds. Rat heart, lung, and kidney before. Decellularization is not new, it was discovered by Poel in 1948 by just preparing acellular mass from muscle tissue [12]. In 1995,

Fig. 3.7 Optical photo and SEM micrograph of one scaffold. Adapted from [10]. Copyright Elsevier, 2010

the first acellular scaffolds were prepared from small intestinal submucosa for the treatment of Achilles tendon [13]. Since this time, numerous decellularized scaffolds have been applied widely in the tissue engineering field.

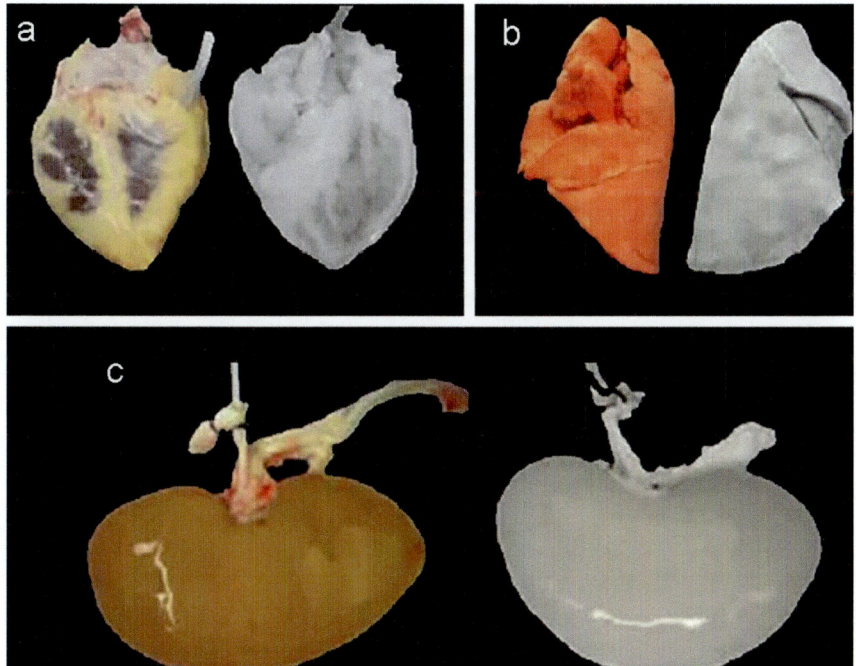

Fig. 3.8 Examples of decellularized organ scaffolds. **a** Rat heart, **b** Rat lung, and **c** Rat kidney before (right) and after (left) decellularization. Adapted from [14]. Copyright Cell Press, 2011

Scaffold prepared by this method possesses superior advantages over polymer-based scaffolds. Where decellularized ECM (dECM) scaffold has a minimum immunological response, it simulates an in vivo microenvironment, and it has biochemical and mechanical properties similar to that of tissues and organs. dECM scaffolds can be classified according to the origin of ECM into autogenous, allogeneic, and xenogeneic dECM. The last two types are common, while autogenous dECM is not widely used due to tissue limitations. There are three methods to prepare dECM scaffolds; physical, chemical, and enzymatic methods. An excellent preparation method depends on the complete elimination of cellular components while maintaining the original architecture, composition, mechanical, and biochemical properties of ECM.

The physical treatment method uses temperature, pressure, and/or force to ease diffusion of detergent to destroy cell membrane and remove cell components subsequently. The common methods used in dECM scaffold fabrication are; immersion and agitation, freeze–thaw cycles, and perfusion. Immersion and agitation treatment rout submerses tissues into decellularization solutions under constant. This method is suitable for fragile, small, and thin sections of organs tissues with simple vascular structures. Freeze–thaw cycles include repeated nitrogen freeze and subsequent thawing in buffer fluid. It is characterized by the preservation of tissue structure and biochemical composition. The perfusion treatment method removes the cell matter by cannulating tissues and organs and then circulates decellularized agents through their inherent vascular system. It is suitable for thicker, larger tissues, or whole organs.

The chemical treatment method is based on the utilization of chemical agents and detergents to destroy cellular bonds to detach cellular components. Detergents used for decellularization are mainly ionic (such as sodium deoxycholate, sodium dodecyl sulfate, sodium lauryl glutamate, sodium lauryl sulfate, sodium lauryl ester sulfate etc.), non-ionic, and zwitterionic. While, the chemical method utilizes acid (e.g. acetic acid, peracetic acid, and hydrochloric acid) and base (e.g. ammonium hydroxide and sodium hydroxide) to dissolve cellular components similar to detergents.

The enzymatic treatment method uses certain types of enzymes, such as nuclease, trypsin, and phospholipase, to specifically break cellular chains into small fragments. The advantage of this method is its selectivity to disintegrate definite cell and ECM components. However, the main disadvantages are acute alteration of ECM structure and in some cases, it is difficult to remove enzymes from the medium.

3.2 Rapid Prototyping (RP)

Rapid prototyping (RP), also known as additive manufacturing (AM), 3D printing (3DP), tool-less manufacturing, and solid freeform (SFF), techniques are computer-aided techniques [15]. It is extensively and progressively has many applications, such as architecture, automation, electronics, aerospace, and biomedical (e.g. tissue engineering scaffold and prosthetic devices) applications. RP techniques fabricate the precise object directly without special tools using the computer and computer-aided design (CAD), where computer software slices the object into layers, and so the object is built layer-by-layer thanks to the movement of the printing head in the z-axis direction. Moreover, the materials, including ceramic, polymer, wood, and metal, used in this technique are nearly unlimited [16]. RP techniques include several types which are based on the working principle of processing, but it can subdivided into three main types, as shown in Fig. 3.9 [17]:

(1) Powder-based type
(2) Photosensitive-based type
(3) Extrusion-based type.

These types are subdivided into categories depending on the concept of machine working. Meanwhile, stereolithography (SLA), selective laser sintering (SLS), fused deposition modeling, and three-dimensional printing (3DP) are the most widely used RP techniques in different applications, generally, and tissue engineering applications, specifically. Tissue engineering scaffold fabricated by RP technique is precise with tailored pore size and shape, good mechanical strength, complex shape can be obtained, and it can be fabricated in a microscale. As well as living cells and biomolecules can be incorporated into scaffold material.

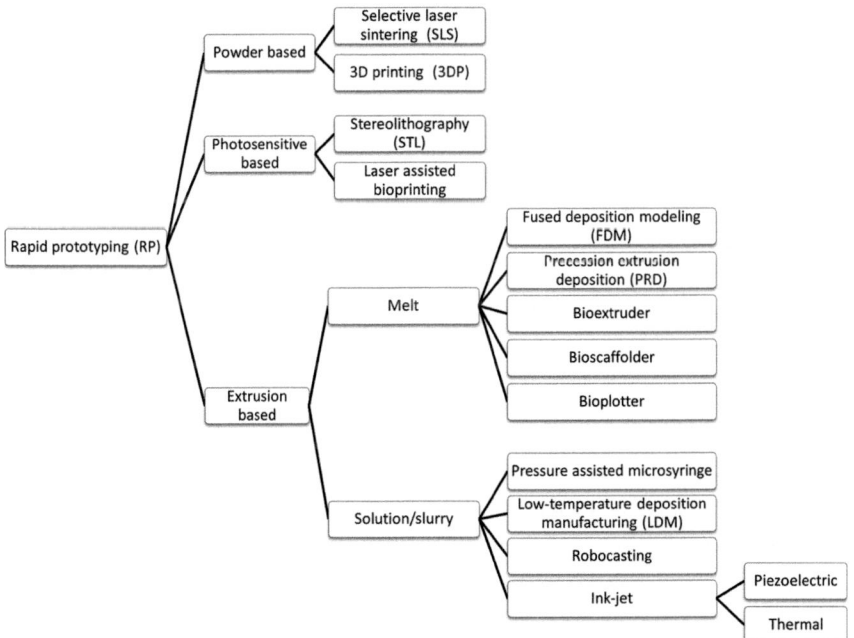

Fig. 3.9 Classification of rapid prototyping (RP) techniques used in tissue engineering application [17]

3.2.1   Stereolithography (SLA)

The stereolithography (SLA) technique is one of the photosensitive-based types. It is based on selective polymerization of photopolymer by UV or laser light. As presented in Fig. 3.10, the device is composed of a photopolymerized monomers (resin) tank, a source of UV or laser light, a dynamic mirror system, 3-step motors for linear motion, and a build platform. Firstly, the tank is filled with resin, and computer software slices the object into layers, each layer can be printed in the x–y plane by controlled movement of the mirror in the x and y directions. The monomers are polymerized by UV or laser light, and so when the first layer is consolidated, the build platform moves down in the z-direction, and the adjacent layer is printed by the same method and so on until completing printing of the whole object. Afterward, the object is collected and washed from uncured resin and subjected to heat treatment or put into a chamber with UV light lamp to complete polymer curing. The main disadvantages of SLA are; that it is not only limited to polymer but especially photopolymer, there is shrinkage in the final printed object, and the starting materials are relatively expensive.

Gabrieliet et al., compared the mechanical strength of two scaffolds with different unit cells, Schwarz primitive and gyroid, and two wall sizes prepared from Dental LT clear resin by stereolithography technique (Fig. 3.11). Their results showed that Young's Modulus increased as the wall thickness increased for the same model. Meanwhile, the mechanical properties of the scaffold of the gyroid unit cell with a length of 4 mm were larger than that of the Schwarz primitive unit cell of the same length [18].

3.2.2   Selective Laser Sintering (SLS)

Selective laser sintering (SLS) is one of the powder-based types in RP technique. It mainly used different types of materials (polymer, ceramic, and metal) in a powder shape through the printing process. SLS is mainly composed of a material reservoir bed (one or two beds) and a powder printing bed. As shown in Fig. 3.10, the role moves toward the printing bed to fill it with a thin layer of material powder. Laser beam selectively sinters this layer to fuse powder particles based on CAD design, after this, the printing bed moves down one step and the powder bed moves up one step also. Likewise, the role moves towards the printing bed to fill it with a new thin powder layer, and then the selective laser sintering process is started again. The process is repeated till the completion of object printing. The advantages of SLS are the ability to use a wide variety of materials, it does not need support material, it can be used to fabricate complex shapes, and it is suitable for scaffolds used for bear loading applications. But, this technique has some limitations, such as the high temperature of the sintering process is not preferred by cells and biomolecules, the printing process needs an inert atmosphere in specific cases, and sometimes fabricated scaffolds need a post-treatment step. Li et al., prepared the diamond model's structure

Fig. 3.10 Schematic diagram of stereolithography (SLA), selective laser sintering (SLS), and fused deposition modeling device setup. Adapted from [19] which is an open access licensed under a Creative Commons Attribution 4.0 International License

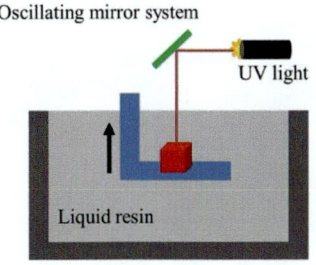

Stereolithography

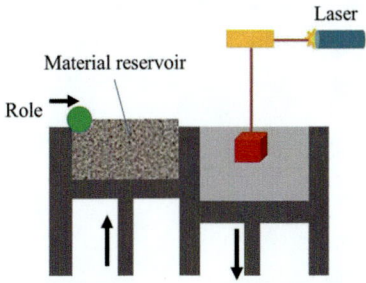

Selective laser sintering

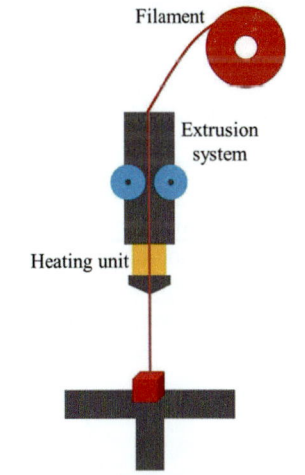

Fused deposition modeling

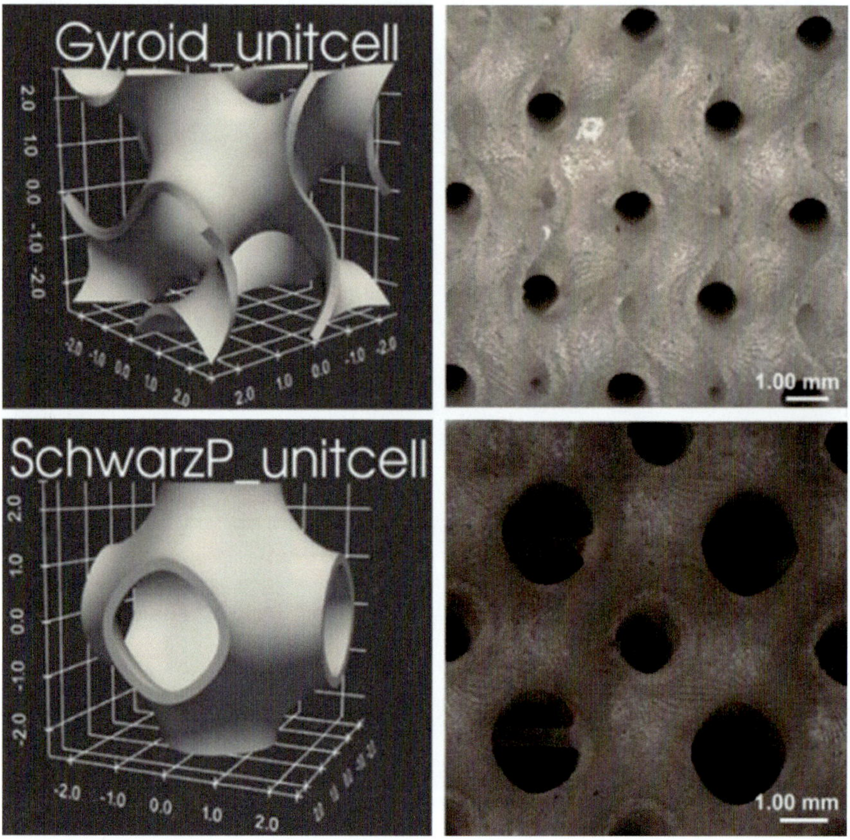

Fig. 3.11 3D design (left) and optical microscopy images (right) of Schwarz primitive type and gyroid type solid unit cells [18]. Copyright MDPI, this article is an open access licensed under a Creative Commons Attribution 4.0 International License

scaffolds based on PLA and magnesium with different weight % (0, 1, 3, 5, and 7%) using the SLS technique (Fig. 3.12). The results showed that the incorporation of magnesium increased the degradation rate of PLA. Moreover, it increased the compressive strength up to 3% magnesium, and it decreased again by increasing magnesium percentage [20].

Fig. 3.12 PLA scaffolds with different ratios of magnesium are prepared by SLS technique. Adapted from [20]. Copyright Springer Nature, 2024

3.2.3 Fused Deposition Modeling (FDM)

Fused deposition modeling (FDM) is one of the extrusion-based types. In this technique, material in a filament shape melted and deposited to build the object as successive layers. The materials usually used are thermoplastic polymers, such as PLA, PEI (polyetherimide), and ABS (acrylonitrile–butadiene–styrene). As presented in Fig. 3.10, the filament is pushed from the filament role into the pre-heated head which melts it, and then the melt is extruded from the nozzle and deposited on the build platform to form the object in a layer-by-layer scenario. In this technique, more than one material can be used by increasing the head numbers. Scaffold prepared by FDM technique is characterized by low shrinkage, relatively high mechanical strength, cost-effective, easy to fabricate, and it does not need a post-treatment process after fabrication. However, the main drawbacks of FDM are the high-temperature printing process which is not suitable for incorporation of cells and biomolecules, it is limited to specific polymers, and the diameter of polymer filament go through nozzle is fixed. Zarei et al., fabricated $CaCO_3$/PLA scaffold by FDM technique, and they subsequently treated the scaffold with cold atmospheric plasma together with the incorporation of $CaCO_3$ (Fig. 3.13) to improve water absorption, roughness, mechanical properties, degradation, and biocompatibility of the scaffold. The results showed that the addition of $CaCO_3$ increased water absorption and roughness, and plasma treatment progressively increased both of them. Likewise, $CaCO_3$ and plasma treatment significantly improved the mechanical properties (tensile modulus and ultimate tensile strength), and increased the degradation rate [21].

3.2.4 Three-Dimensional Printing (3DP)

3D printing (3DP) technique is closely analogous to SLS technique, but the solution (inkjet) is used to bind powder particles instead of the laser beam. Likewise, a wide variety of materials can be applied in this technique, such as ceramics, polymers, metals, and a combination of two materials. The main advantageous property compared to SLS is the cold printing process which is appropriate to combine cells and biomolecules during the printing process. The most potential application of 3DP is bioprinting. In which a combination of cells, growth factors, biomolecules and biodegradable polymer (hydrogel) can be used to fabricate living tissue directly. Ex vivo tissue, organoids, microenvironments, and organ-on-chip can be fabricated by 3D bioprinting. The most commonly used methods in bioprinting are inkjet printing, laser-assisted, and extrusion.

Inkjet printing is also known as drop-on-demand bioprinting because bioink ejects as drops in a controlled manner. It is divided into piezoelectric inkjet, thermal inkjet, and electrostatic inkjet. Laser-assisted bioprinting uses a long-wavelength laser beam with a controlled pulsating rate to print cells on the receiving substrate. Extrusion bioprinting is based on the extrusion of printed material by mechanical

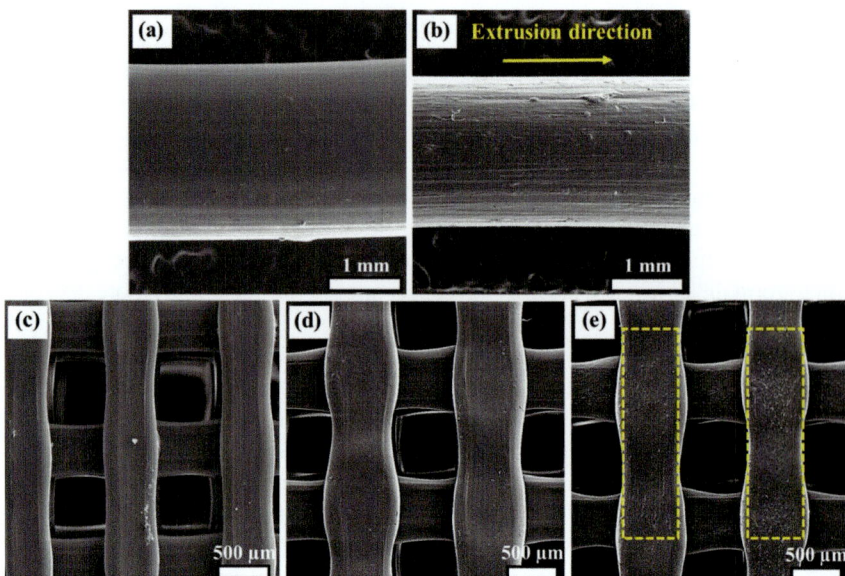

Fig. 3.13 SEM micrographs of **a** PLA and **b** PLA-CaCO₃ filaments. SEM micrographs of **c** PLA scaffold, **d** PLA-CaCO₃ scaffold, and **e** PT-PLA-CaCO₃ scaffold. PT is plasma-treated [21]. Copyright Elsevier, 2024

force which is a pneumatic pump, screw or piston, or air pressure force (Fig. 3.14) [22]. Yu et al., fabricated a biphasic gelatin–methacryloyl scaffold by 3D bioprinting (Fig. 3.15). The first phase was loaded with bone marrow mesenchymal stem cells (BMSCs) and articular chondrocytes which mimic cartilage layer, while the second phase was gelatin-methacryloyl/Sr-substituted xonotlite loaded with bone marrow mesenchymal stem cells which mimic the subchondral bone layer and stimulated differentiation of BMSCs into osteoblasts. They carried out in vitro chondrogenesis and osteogenesis tests, as well as, in vivo evaluated the scaffolds by implanting them in rat osteochondral defects of diameter 2 mm and depth 2 mm. The results demonstrated that the scaffolds stimulated cell differentiation, and they fully regenerated osteochondral defects in the rats [23].

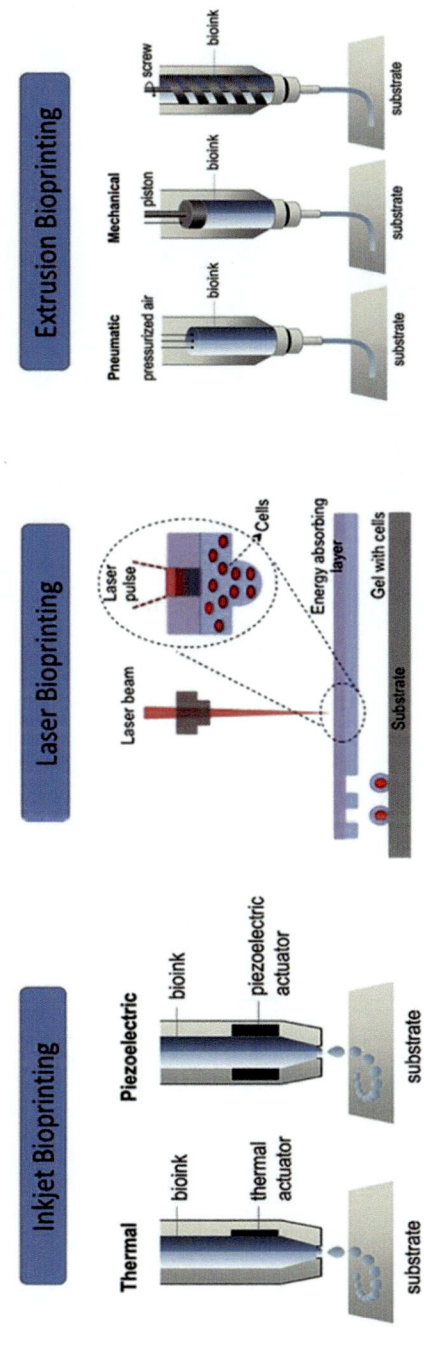

Fig. 3.14 Common used bioprinting techniques: inkjet, laser bioprinting, extrusion bioprinting. Adapted from [22]. Copyright Elsevier, 2024

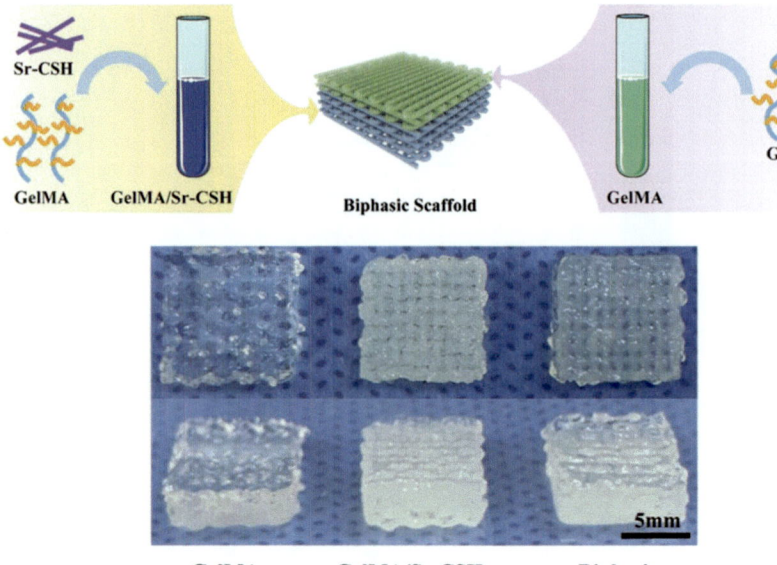

Fig. 3.15 Biphasic multicellular gelatin-methacryloyl scaffolds. Schematic presentation of two bioinks corresponding to two phases used in the scaffold fabrication (upper). As-prepared scaffolds (lower) [23]. John Wiley and Sons, this an open access article distributed under the terms of the Creative Commons CC BY license

References

1. Roseti L, Parisi V, Petretta M, Cavallo C, Desando G, Bartolotti I et al (2017) Scaffolds for bone tissue engineering: state of the art and new perspectives. Mater Sci Eng C 78:1246–1262
2. Hammouda HF, Farag MM, El Deftar MMF, Abdel-Gabbar M, Mohamed BM (2022) Effect of ce-doped bioactive glass/collagen/chitosan nanocomposite scaffolds on the cell morphology and proliferation of rabbit's bone marrow mesenchymal stem cells-derived osteogenic cells. J Genet Eng Biotechnol 20:33
3. Hutmacher DW (2000) Scaffolds in tissue engineering bone and cartilage. Biomaterials 21:2529–2543
4. Thadavirul N, Pavasant P, Supaphol P (2014) Development of polycaprolactone porous scaffolds by combining solvent casting, particulate leaching, and polymer leaching techniques for bone tissue engineering. J Biomed Mater Res Part A 102:3379–3392
5. Soundarya SP, Menon AH, Chandran SV, Selvamurugan N (2018) Bone tissue engineering: scaffold preparation using chitosan and other biomaterials with different design and fabrication techniques. Int J Biol Macromol 119:1228–1239
6. Li Z, Xie M-B, Li Y, Ma Y, Li J-S, Dai F-Y (2016) Recent progress in tissue engineering and regenerative medicine. J Biomater Tissue Eng 6:755–766
7. Sun B, Long Y, Zhang H, Li M, Duvail J, Jiang X et al (2014) Advances in three-dimensional nanofibrous macrostructures via electrospinning. Prog Polym Sci 39:862–890
8. Lu T, Li Y, Chen T (2013) Techniques for fabrication and construction of three-dimensional scaffolds for tissue engineering. Int J Nanomed 337–350
9. Gautam S, Sharma C, Purohit SD, Singh H, Dinda AK, Potdar PD et al (2021) Gelatin-polycaprolactone-nanohydroxyapatite electrospun nanocomposite scaffold for bone tissue engineering. Mater Sci Eng C 119:111588

10. El-Kady AM, Saad EA, El-Hady BMA, Farag MM (2010) Synthesis of silicate glass/poly(l-lactide) composite scaffolds by freeze-extraction technique: characterization and in vitro bioactivity evaluation. Ceram Int 36:995–1009

11. Rana D, Zreiqat H, Benkirane-Jessel N, Ramakrishna S, Ramalingam M (2017) Development of decellularized scaffolds for stem cell-driven tissue engineering. J Tissue Eng Regen Med 11:942–965

12. Poel WE (1948) Preparation of acellular homogenates from muscle samples. Science 108:390–391

13. Badylak SF, Tullius R, Kokini K, Shelbourne KD, Klootwyk T, Voytik SL et al (1995) The use of xenogeneic small intestinal submucosa as a biomaterial for Achille's tendon repair in a dog model. J Biomed Mater Res 29:977–985

14. Song JJ, Ott HC (2011) Organ engineering based on decellularized matrix scaffolds. Trends Mol Med 17:424–432

15. Gibson I, Rosen DW, Stucker B, Khorasani M, Rosen D, Stucker B et al (2021) Additive manufacturing technologies. Springer, Berlin

16. Jahed E, Khaledabad MA, Almasi H, Hasanzadeh R (2017) Physicochemical properties of *Carum copticum* essential oil loaded chitosan films containing organic nanoreinforcements. Carbohyd Polym 164:325–338

17. Madrid APM, Vrech SM, Sanchez MA, Rodriguez AP (2019) Advances in additive manufacturing for bone tissue engineering scaffolds. Mater Sci Eng C 100:631–644

18. Gabrieli R, Wenger R, Mazza M, Verné E, Baino F (2024) Design, stereolithographic 3D printing, and characterization of TPMS scaffolds. Materials 17:654

19. Liu K, Hu N, Yu Z, Zhang X, Ma H, Qu H et al (2023) 3D printing and bioprinting in urology. IJB 9

20. Li M, Yuan H, Ding W, Du H, Guo X, Li D et al (2024) Selective laser sintering PLA/Mg composite scaffold with promoted degradation and enhanced mechanical. J Polym Environ

21. Zarei M, Hosseini Nikoo MM, Alizadeh R, Askarinya A (2024) Synergistic effect of $CaCO_3$ addition and in-process cold atmospheric plasma treatment on the surface evolution, mechanical properties, and in-vitro degradation behavior of FDM-printed PLA scaffolds. J Mech Behav Biomed Mater 149:106239

22. Mobaraki M, Ghaffari M, Yazdanpanah A, Luo Y, Mills DK (2020) Bioinks and bioprinting: a focused review. Bioprinting 18:e00080

23. Yu X, Gholipourmalekabadi M, Wang X, Yuan C, Lin K (n/a) Three-dimensional bioprinting biphasic multicellular living scaffold facilitates osteochondral defect regeneration. Interdisc Mater

Chapter 4
Recent Clinical Applications of Biomaterials in Tissue Engineering

Abstract This chapter represents recent applications of biomaterials regenerative medicine, and their evaluation clinically for bone, cartilage, skin, liver, vascular tissue, and skeletal muscles. The function and anatomical composition are reviewed to show the ideal scaffold properties and tissue regeneration conditions required to achieve tissue reconstruction. This presentation will serve to illustrate of challenge facing the application of biomaterials in the treatment of diseased tissue and regenerate tissue in Chap. 5.

4.1 Bone Tissue Regeneration

First of all, before discussing the reconstruction of specific tissue by biomaterials, it is useful to understand the composition and mechanism of formation of this tissue. Bone forms the skeletal parts of the body which supports movement and protects the internal organs (such as the heart, liver, lungs, and brain). Moreover, it acts as blood and fat storage for bone marrow. An environment for marrow (both blood-forming and fat storage) [1]. It is rigid and hard tissue composed of inorganic and organic components. Inorganic components act 69% of the whole bone mass which is mainly composed of hydroxyapatite crystals (99%). Organic components present 22% including collagen (90%) and noncollagen structural proteins. There are two types of bone; cortical bone, solid and trabecular bone. Cortical bone is dense and surrounds trabecular bone which is characterized by a highly interconnected porous structure. The bone tissue possesses a unique structure throughout its macroscale to atomic scale. Where bone tissue is composed of osteons and Harversian canals, each osteon is built by twined collagen molecules forming tropocollagen strands aligned to form repeated districts connected by HAp nanocrystals. A Haversian canal presents in the center of the osteon and it provides the blood to the bone tissue (Fig. 4.1) [2].

The process of bone formation is known as osteogenesis or ossification. Bone is formed in the early stage of development of the organs of the embryo. It starts between the sixth and seventh weeks of embryonic formation, and it continues up to adulthood (about the age of 25 years) [5]. Osteogenesis occurs by intramembranous

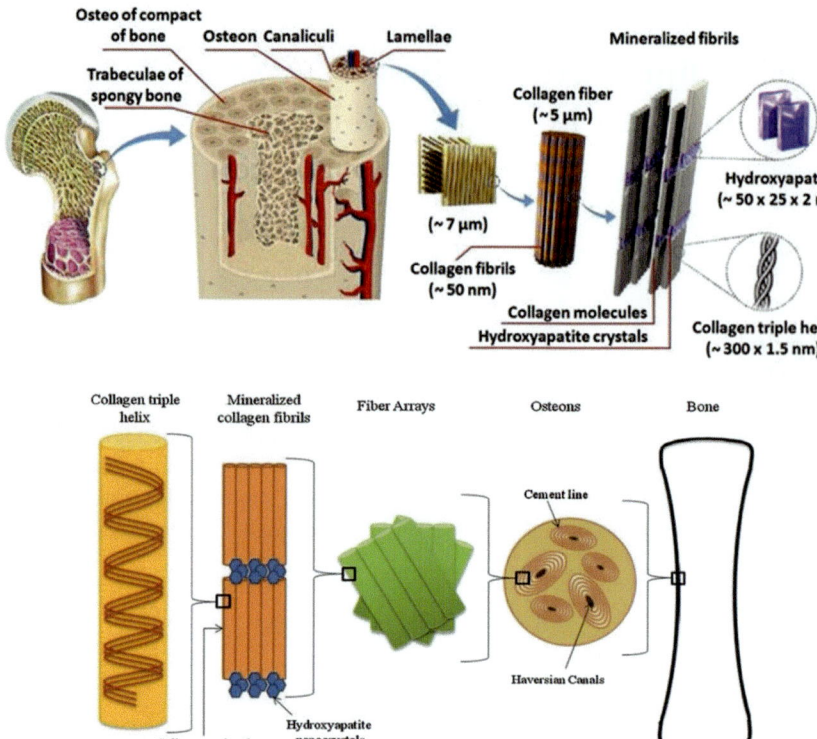

Fig. 4.1 Bone tissue structure (upper) [3], Copyright Springer Nature, 2017. Simplified bone tissue structure (lower) [4], John Wiley and Sons, 2021, this is an open access article distributed under the terms of the Creative Commons CC BY

and endochondral, and three types of cells share in bone development, growth, and remodeling which are osteoblasts (initiate from mesenchymal stem cells), osteocytes (they have finally differentiated osteoblasts and they support bone structure and metabolism), and osteoclasts (they degrade bone to initiate normal bone remodeling). Intramembranous ossification forms bones of the skull, clavicle, and mandible directly from mesenchyme. While, endochondral ossification forms bones (such as radium, femur, and tibia) using cartilage.

Bone defects can result from fracture, bone deformation, trauma, infection disease (osteomyelitis), and osteoporosis. Treatment of bone defects requires reconstructing and regenerating by natural or synthetic transplantation to restore the structure and function of the bone [6]. There are more than 4 million people who require treatment for bone defects in the world [7], and this number is progressively increasing every year due to the increase in population. Therefore, the treatment of bone defects is a critical issue and it possesses great clinical significance. Bone grafting is an extensively used method in the treatment of bone defects. Bone grafts are subdivided into; autograft (bone tissue is taken from the same patient), allograft (bone tissue is

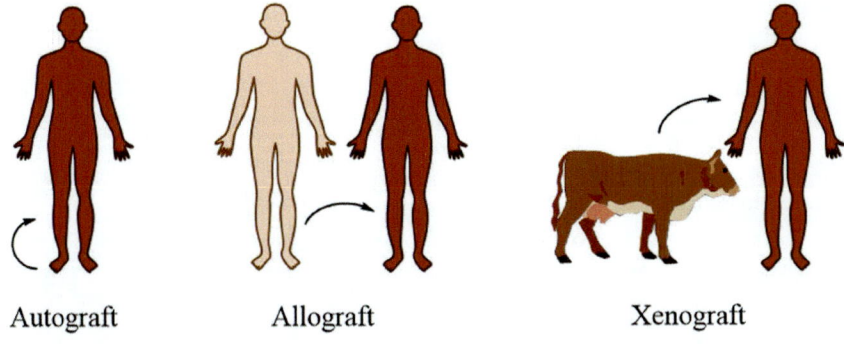

Autograft   Allograft   Xenograft

Fig. 4.2 Types of bone grafting according to the source of bone tissue transplantation

taken from another person), xenograft (bone tissue is taken from another species other than human), and synthetic graft (they are manmade materials). Figure 4.2 shows different types of bone grafting. Autologous bone is considered the gold standard due to its osteo-induction and osteo-conduction with evidenced osteogenesis, but lack of donor tissue limits autografting application [8]. Likewise, bone allografting and xenografting are restricted for bone disease treatment due to immune rejection risks.

Consequently, synthetic bone graft substitutes can compensate limitations of other bone grafting when they achieve osteoinductivity, osteoconductivity, osseointegrativity, high biocompatibility, and high mechanical properties [9]. Synthetic bone grafts can be formulated from synthetic and/or natural biomaterials. As mentioned in Chap. 3 bone grafting materials consist of Ca-phosphate ceramics (e.g. HAp, β-TCP, CDHA, Ca-phosphate cements), glass, glass ceramics, and metal alloys. These materials can be combined with polymer (e.g. collagen, chitosan, gelatin, etc.) to form composite materials with improved properties. As well as fabrication of bone grafting with architecture that mimics the bone structure and provides angiogenesis and vascularization which are important in the delivery of various vital nutrients for osteogenesis activity is an important issue. Scaffold used in bone tissue engineering is an ideal substrate that mimics the bone composition and architecture. Numerous studies stated that the pore size of bone engineering scaffolds should be $\geq 300\ \mu m$ for cell migration, cell proliferation, and osteogenesis [10]. Where, Wu et al. confirmed in their study that Mg-substituted wollastonite scaffolds, fabricated by stereolithography technique, of 450 and 600 μm improved the formation and remodeling of new bone tissues [11]. We prepared ceramic scaffolds with different pore sizes and architecture based on Mg-phosphate at room temperature by a combination of 3D printer paste extrusion deposition technique and ceramic cementation concept (Fig. 4.3). Where the scaffold green body was prepared first, and then it was immersed in ammonium phosphate dibasic as a cement solution to convert it to a struvite-based scaffold. Lysozyme drug was loaded into the scaffolds thanks to the fabrication process being carried out at room temperature. The scaffolds showed desirable mechanical strength and good viability with MC3T3-E1 cell line [12]. In an

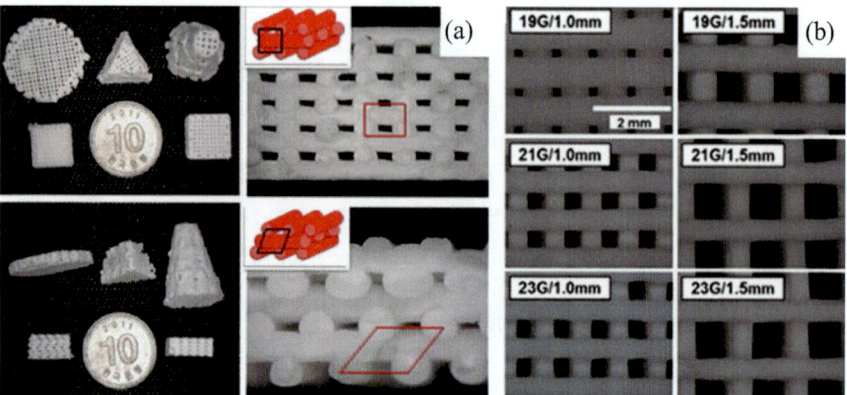

Fig. 4.3 Different MgP scaffold shapes built of center-to-center distances between struts (**a**), and scaffolds fabricated with different nozzle sizes, 21G and 23G (**b**) [12]. Copyright Elsevier, 2014

extensive study, we prepared Mg-phosphate/gelatin composite scaffolds following the previous method, and the paste viscosity was controlled by temperature during the rapid prototyping process. The results concluded that gelatin clearly improved the scaffold elastic modulus, and enhanced MC3T3-E1 cell attachment [13].

4.2 Cartilage Tissue Regeneration

Cartilage is a flexible connective tissue that provides a low-friction gliding surface and absorbs impact during body movements. It presents in joints, nose, intervertebral discs, and ear. Cartilage starts in the embryo from mesenchyme cells which aggregate to form blastema cells, this occurs in the fifth week of pregnancy. The formed blastema starts to form chondroblasts. Chondrocytes are formed separately thereafter in the extracellular matrix. Cartilage is an aneural, and alymphatic, avascular (which leads to hypoxic environments) structure [14]. It is composed of water (65–80%), collagen, mainly Type II, (10–20%), proteoglycans which are protein polysaccharide molecules that impart compressive strength to the cartilage, and chondrocytes which are highly specialized cells sparingly spread within the matrix, they synthesize all the matrix components and regulate matrix metabolism [15]. Cartilage can be divided into three zones, superficial zone, middle zone, and deep zone (Fig. 4.4). The superficial zone is the outer cartilage surface, it presents 10–20% of the cartilage thickness, and it mainly contains type II collagen fibers flat and ellipsoidal chondrocytes parallel to the surface cartilage surface. The middle zone acts 40–60% of the total cartilage thickness, and it consists of randomly distributed round or rectangular chondrocytes perpendicular to the cartilage surface. In addition, type II collagen fibers appear as oblique arcades. The deep zone shows 20–30% of the cartilage thickness. It includes

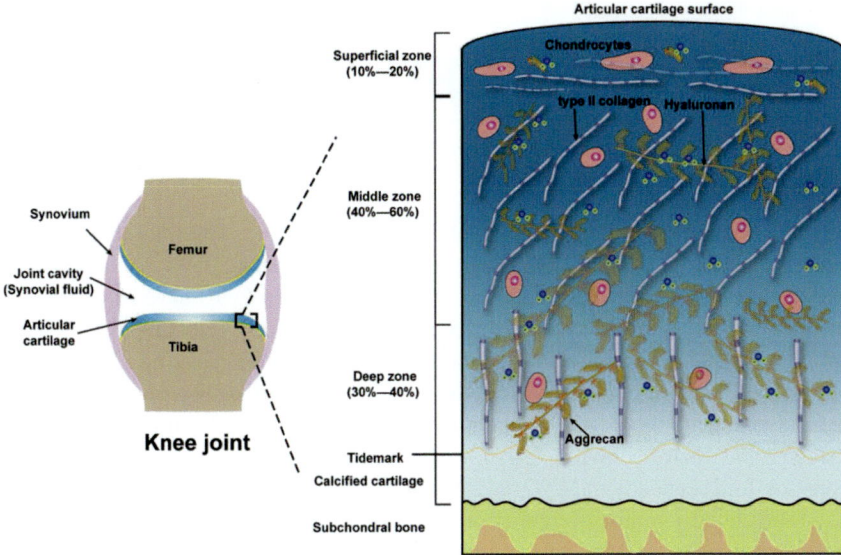

Fig. 4.4 Schematic diagram of the knee joint (left) shows the cavity filled with lubricant (synovial fluid), and articular cartilage (right) presents different zones of cartilage with characteristic components [16]. This article is an open access licensed under a Creative Commons Attribution 4.0 International License

rounded chondrocytes are round and type II collagen fibers perpendicular to the surface of cartilage. The interfacial zone is in contact with the lubricant present in the joints, while the deep zone is attached to the bone tissue.

Due to its avascular structure and made up of specialized cells (chondrocytes), cartilage regeneration is limited, and so, repair and reconstruction of damaged cartilage tissue is a crucial issue. Meanwhile, the treatment method of cartilage damage depends on the place, area, and depth of damage, as well as, age, patient physical activity, and chronicity. The depth and area are the most criteria that describe the cartilage damage degree which is known as the Outerbridge classification [17]. In this articular cartilage lesions are classified into five grades (Fig. 4.5); Grade 0 is a soft surface, normal, and healthy cartilage. Grade I; in which the cartilage surface is soft and swollen with increased water content. Grade II; in this case, cracks are propagated into the cartilage surface and the crack length reaches half the thickness of the cartilage. Grade III; herein the depth of cartilage defect is more than half the thickness of the cartilage. Grade IV; the defect increases to the degree that the subchondral bone is completely uncovered with the cartilage. The cell-based method is considered the most promising method for cartilage treatment approaches. Where chondrocytes can be taken from a healthy cartilage present in the patient himself. This process requires 3D scaffold is essential as a temporary substrate for chondrocytes.

Wang et al. implanted gelatin methacrylamide, ε-poly-L-lysine, and/or 3-Aminophenylboronic acid hydrogel seeded with chondrocytes subcutaneously into

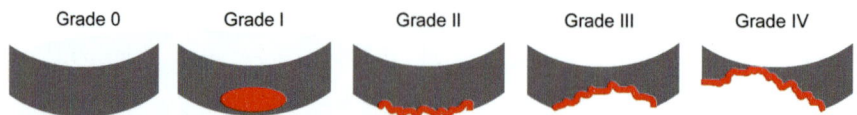

Grade 0   Grade I   Grade II   Grade III   Grade IV

Fig. 4.5 Outerbridge classification of articular cartilage defects

the back of a rat. The results showed that the hydrogel supported the tissue regeneration of cartilage defects [18]. Meanwhile, several types of cells (such as widely used mesenchymal stem cells (MSCs)) other than chondrocytes have been used for cartilage regeneration. Li et al., fabricated PCL scaffold by electrospinning technique and studied its ability to enhance in vitro chondrogenesis of MSCs in the presence of transforming growth factor-β (TGF-β). Their results showed that the scaffold enhanced the degree of chondrogenesis of MSCs [19]. Multichamber flow perfusion bioreactor shown in Fig. 4.6 was used to study the ability of PCL nanofibrous scaffold seeded with MSCs to form an extracellular matrix (ECM) of cartilage tissue and chondrogenic differentiation. Such a reactor improved the ECM formation and differentiation of MSCs to chondrocytes [20]. Heirani-Tabasi et al. fabricated injectable hydrogel based on chitosan/hyaluronic acid seeded with adipose-derived MSCs and chondrocyte extracellular vesicles. The results showed that the expression of chondrogenic genes was enhanced by an addition of chondrocyte extracellular vesicles [21]. Haghighi et al. prepared scaffolds of pore size range 100–300 μm based on gelatin/chitosan/silk fibroin, and the optimum scaffold seeded with human chondrocytes and in vivo examined in the rabbit and compared with unseeded one. The results exhibited the development of new cartilage tissue by a cell-containing scaffold greater than that of cell-free one [22].

Recently, decellularized cartilage ECM combined with other materials to fabricate scaffolds used for cartilage reconstruction. The decellularization process (as presented in Chap. 3) can be carried out by physical, chemical, and enzymatic

Fig. 4.6 Multichamber flow perfusion bioreactor used in culturing of PCL nanofibrous scaffold seeded with MSCs [20]. Copyright American Chemical Society, 2010

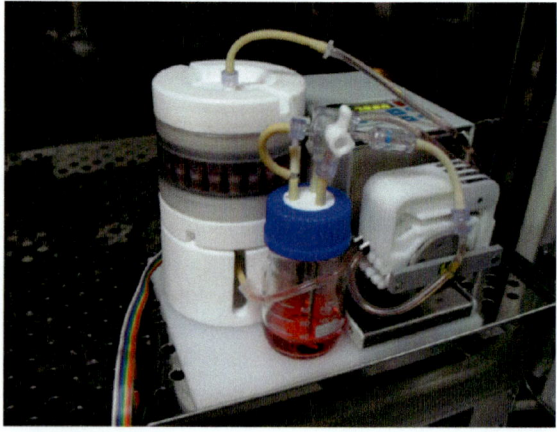

methods. Physical decellularization is performed by freeze-thawing, osmotic pressure, sonication, hydrostatic pressure, or electroporation [23]. The most widely used method is freeze-thawing which is carried out repeatedly, where the cell membranes are perforated by ice crystals. Despite this method being considered the least disruptive method for the ECM structure and components it is not completely the cell remains and genetic debris. Chemical decellularization is more effective and versatile than physical one. It can be carried out by acid, base, or detergents (commonly used types; ionic, nonionic, and zwitterionic). Acids, such as formic acid and peracetic acid, are commonly used in chemical decellularization, based on the removal of the cellular components. While detergent depends on denaturing the cell membranes and proteins. Enzymatic decellularization is usually used after chemical decellularization to remove nuclear material remains. The most widely used enzymes are nucleases (they work on DNA and RNA chains) and proteases (act on peptide bonds).

Using of decellularized cartilage ECM is considered a recent trend for cartilage regeneration. Stone et al., prepared ECM scaffold from decellularized porcine auricular cartilage for chondrocytes attachment and growth. Figure 4.7 shows an example of decellularized cartilage of porcine auricular cartilage. They deeply made biomarker analyses to measure the remaining materials after decellularization. The histological and molecular results showed that the resulting ECM scaffold contained a very small quantity of cells remainings [24]. Yaqiang et al. fabricated a scaffold based on bacterial cellulose and decellularized cartilage ECM for cartilage reconstruction. The scaffolds were characterized by excellent mechanical strength, high water uptake, and shape-memory properties. Moreover, the scaffolds showed good cell viability, and they in vivo enhanced the formation of neocartilage and tissue regeneration [25]. Ayariga et al., prepared a decellularized avian (*Gallus gallus domesticus*) cartilage scaffold and reseeded it with human chondrocytes. The scaffold demonstrated stiffness similar to that of cartilage tissues. Moreover, the scaffold was cytocompatible and did not show an immunological response [26].

4.3 Skin Tissue Regeneration

The skin is the largest organ of the body and plays an important role in protecting the body from water, temperature, infection by microorganisms, ultraviolet light, and mechanical and chemical shock. It is composed of three main layers from outside to inside; the epidermis, the dermis, and the hypodermis (subcutaneous) layer (Fig. 4.8). The outermost portion of the epidermis layer is the Stratum corneum which is relatively less hydrophilic than the inner layers. It includes keratinocytes, melanocytes (which are responsible for skin color by producing melanin pigment and filtering the UV radiation), and Langerhans cells (share in the immune system of the skin). The dermis layer is the middle layer which has the largest thickness relative to the other two layers. It is composed of extracellular matrix (ECM), vascular endothelial cells, fibroblasts (secretes elastin and collagen which provides skin elasticity), hair

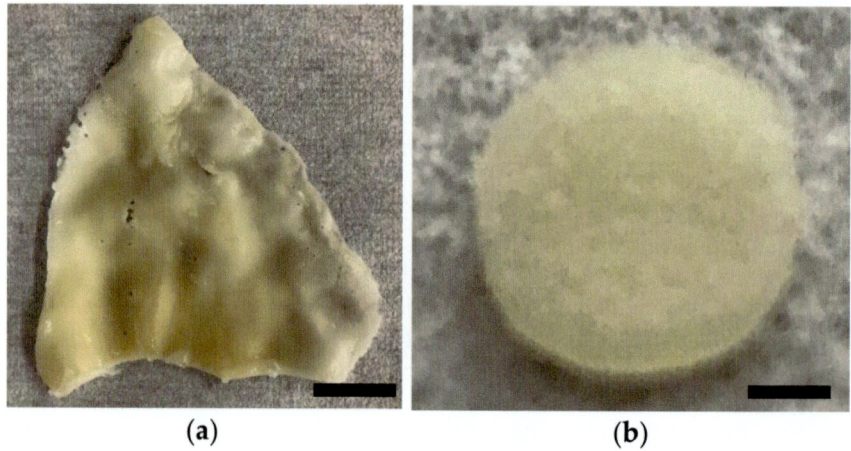

follicles, blood vessels, sweat glands, subcutaneous glands, and nerve endings [27]. The hypodermis layer which is mainly adipose tissue acts as energy storage, and it acts as a barrier between the skin and bone and muscles [27].

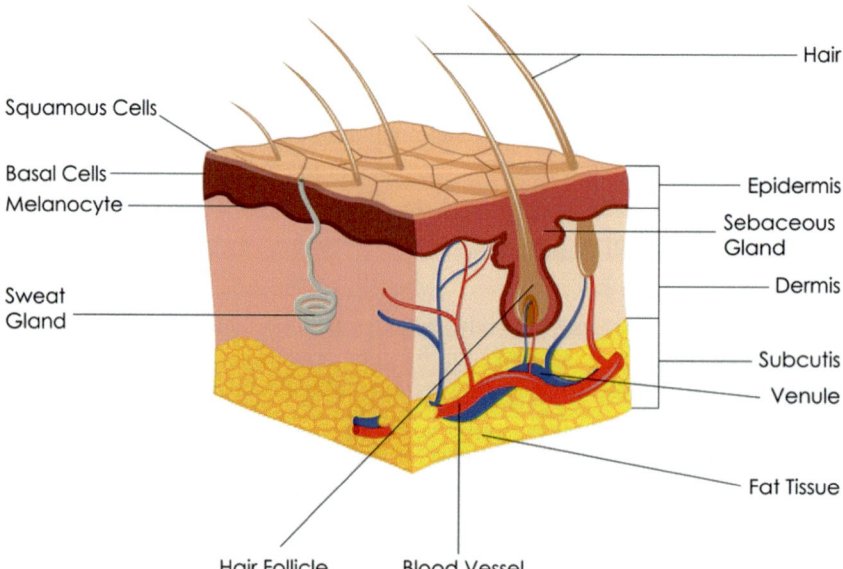

Fig. 4.8 Skin structure and composition [29]

The route of skin wound healing after injury occurs by cutaneous response which proceeds in three successive steps; inflammation, proliferation, and maturation (Fig. 4.9). In the inflammation step, a temporary shield is formed by the coagulation of blood which protects the body from microorganisms invasion. Also, the blood flow increases at the wound zone and forms a fibrin matrix which is occupied by immune cells (monocytes and neutrophils) to remove the dead tissue and start regeneration of new tissue. This step takes about 4 days [28]. The next step is proliferation, which occurs during 5–20 days. Herein, the endothelial cells and fibroblasts are stimulated, and collagen secreted by fibroblasts gradually replaces the fibrin matrix. The wound area is reduced actin released by myofibroblasts which originated from fibroblasts. Then, the angiogenesis begins causing the development of new granulation tissue, and keratinocytes migrate to the surface of this tissue underneath the blood clot [28]. The maturation step can take a long time to reach to years. Where wound re-epithelialization occurs in the dermis [27].

Small wounds are treated usually with traditional wound dressings including gauze and tulle. Meanwhile, chronic and larger wounds require potential methods to induce skin tissue regeneration without leaving a trace on the skin. In addition, skin grafting and replacing large parts of damaged skin resulting from accidents, burns, and diabetes with new skin tissue is still attracting interest of the scientists. Scaffold substrate is widely used for skin tissue regeneration. Moreover, it can be functionalized with antibacterial molecules to prevent bacterial infection during the healing process. Nanofibrous scaffolds made by electrospinning technique are widely used for wound dressing. Sarkar et al. synthesized chitosan/collagen electrospun

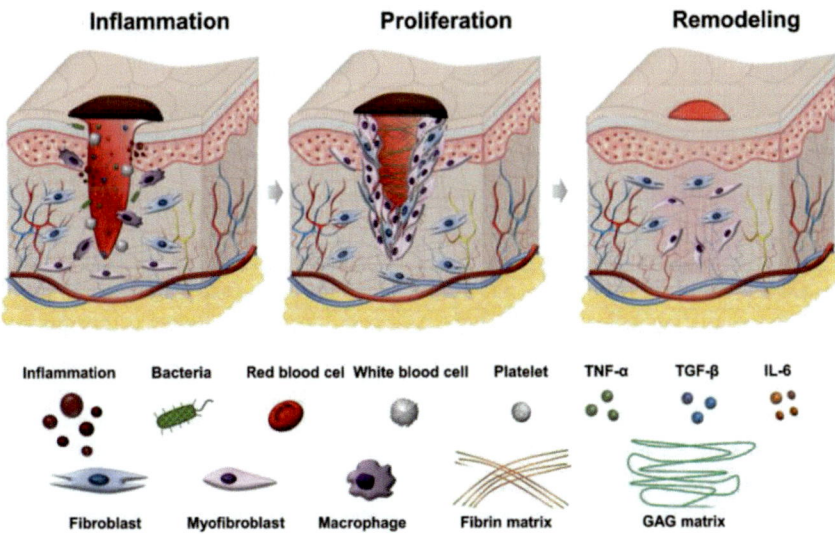

Fig. 4.9 Phases of wound healing, inflammation, proliferation, and maturation phase (after modification from [30]). MDPI, 2022. This article is an open access licensed under a Creative Commons Attribution 4.0 International License

mats for skin tissue engineering applications. They first made chitosan fibrous mats and then immersed them in collagen polymer solution followed by cross-linking of chitosan and collagen. They studied the viability of prepared scaffold against HaCaT keratinocyte and 3T3 fibroblast cells, and evaluated scaffold wound healing ability on an ex vivo human skin equivalent (HSE) wound model (they are human skin samples collected from patients). The results displayed the detection of keratinocyte migration and re-epithelization of wound which are evidence of wound healing [31]. Khalilimofrad et al., also prepared gelatin/chitosan electrospun mats but cross-linked to collagen type I thereafter. Cell viability tests with L929 (NCBI C161) cell line revealed that collagen type I enhanced cell growth and migration [32].

Seeding of a scaffold with cells (such as mesenchymal stem cells, stem cells, epithelial cells, and adipose stem cells) has improved the scaffold for skin regeneration. For example, Sun et al. incorporated adipose-derived mesenchymal stromal cells in different electrospun scaffolds based on PLA, fibrinogen, or collagen. They in vivo studied the wound healing capability of the scaffolds in skin defects of diabetic rats. Their findings exhibited that the scaffolds improved wound healing [33]. In a similar work, Xu et al., prepared an antibacterial bioactive electrospun wound dressing based on PLA/PVA-contained silver nanoparticles. The results showed that the prepared scaffolds stimulated extracellular matrix synthesis and showed anti-inflammatory properties while promoting angiogenesis, and they in vivo enhanced the wound healing process as shown in Fig. 4.10 [34]. Pazhouhnia et al., prepared a bi-layered gelatin-methacryloyl (GelMA)/gelatin scaffold by 3D bioprinting. The upper layer included keratinocytes, and the lower section contained fibroblasts and human umbilical vein endothelial cells (HUVEC) cells to mimic the skin structure. The results displayed that the addition of gelatin to GelMA improved the scaffold mechanical properties. Overall, the scaffold induced formation of extracellular matrix and well-defined skin layers [35]. Girard et al., synthesized PCL scaffold/membrane bilayer to mimic natural skin structure. The membrane was PCL solution electrospun layer, while the scaffold was made of PCL by a melt electrowritten method. Such scaffold was seeded with human fibroblast and keratinocyte cells. They stated that this bilayer model can be used for testing new pharmaceutics and cosmetics applied to the skin. Moreover, it is considered a versatile tool for the evaluation of wound dressing material from the view of wound healing and skin regeneration [36].

Recently, electricity can be used to stimulate wound healing using electric current [37]. The stimulating electricity (SE) can be alternating current, direct current, pulsed current, low-intensity direct current, and high-voltage pulsed current, each modality is suitable for a specific wound (Fig. 4.11). Electrical stimulation serves wound healing during its three phases; inflammation (it increases blood flow, stimulates fibroblasts, and increases tissue oxygenation and antibacterial activity), proliferation (increases wound contraction, collagen matrix organization, membrane transport, and the stimulation of protein and DNA synthesis), and remodeling (it stimulates fibroblasts and enhances epidermal cell proliferation and migration). Furthermore, SE can be useful in controlling the release of therapeutic compounds (e.g. antibacterial drugs and metal oxides) loaded on the scaffold and wound dressing materials. Conductive materials are beneficial on this side when they are incorporated in the wound dressing

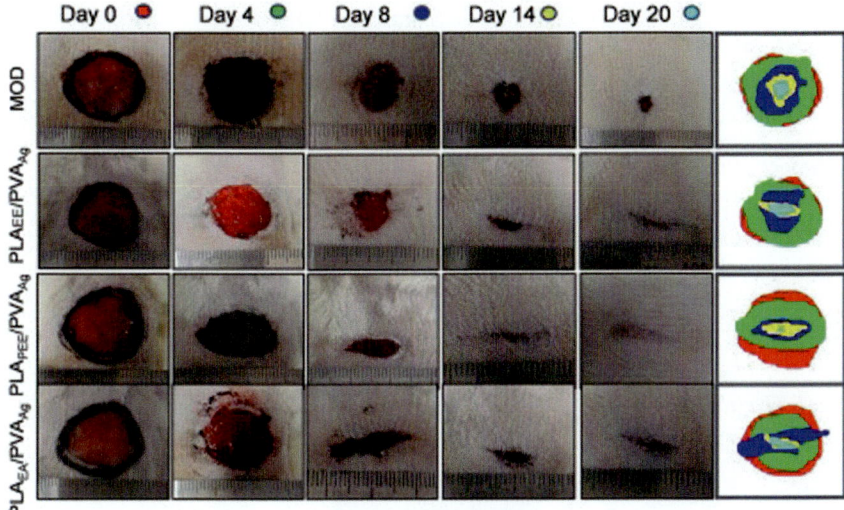

Fig. 4.10 Effect of different scaffolds on the wound healing of skin rats. Adapted from [34]. Copyright Elsevier, 2024

materials because they facilitate electric current transportation. Conductive materials can be polymers (e.g. polyaniline, polypyrrole, and polythiophene), metals, carbon nanotubes, graphene oxide, metals, and metal oxides. Liu et al., prepared conductive hydrogel containing black phosphorus which possessed the ability to transform from a sol state to a gel state under electric current. Black phosphorus did not only show in this study good electrical conductivity and promotion of wound healing but it also presented antibacterial activity [38]. Aycan et al., prepared electroconductive and anti-inflammatory composites by incorporating graphene oxide and ibuprofen in sodium alginate/hyaluronic acid/gelatin matrix for wound dressing. The results exhibited that high conductive composites improved cell attachment and proliferation, as well as, the substrate loaded with ibuprofen showed a good anti-inflammatory effect [39]. Najafian et al., synthesized electroconductive polyaniline-grafted tragacanth gum/poly(vinyl alcohol) scaffolds by electrospinning technique for skin tissue engineering application. The electrical conductivity of the scaffolds was measured by the four-probe technique. Moreover, the cell viability of the scaffolds was studied against mouse fibroblast L929 cells. The results showed that scaffolds incorporation of tragacanth gum decreased the electroconductivity of the scaffold. On the other hand, the scaffolds enhanced cell adhesion and proliferation [40].

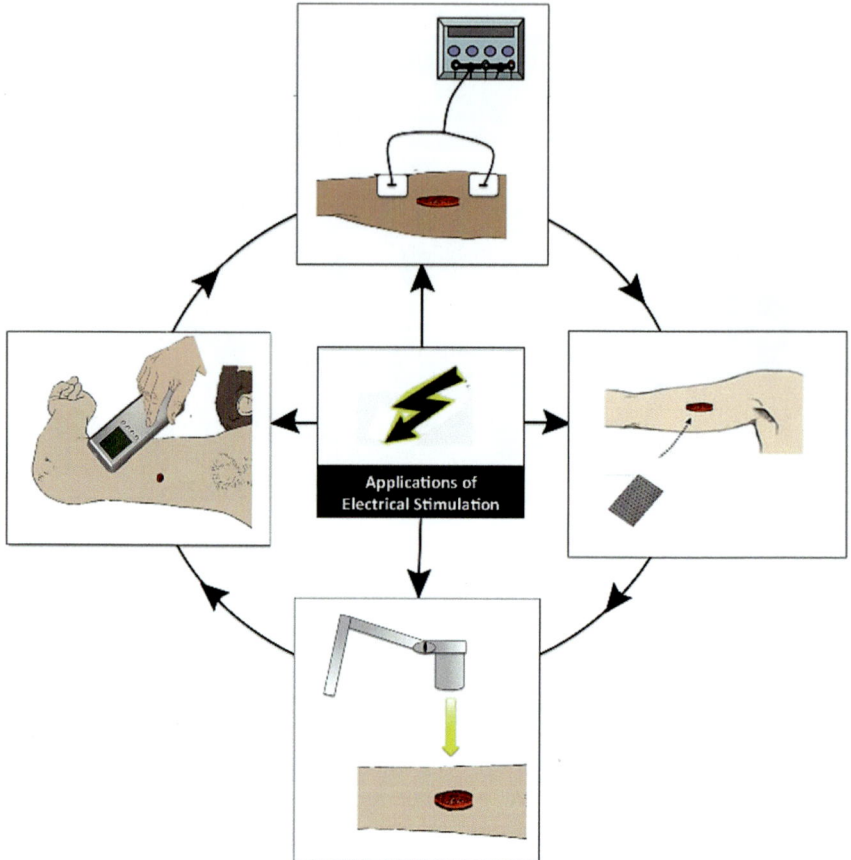

Fig. 4.11 Stimulation of wound healing by using different routes of electricity [41]. MDPI 2014, this article is an open access licensed under a Creative Commons Attribution 4.0 International License

4.4 Liver Tissue Regeneration

The liver is considered one of the complex organs according to the microarchitecture, and it is the largest solid organ in the body. The liver has vital functions, such as albumen and bile production, blood filtration, regulation of amino acids and blood clotting, removal of bacteria and excess glucose from the blood, and storage of minerals (Cu and Fe) and vitamins (A, D, E, K, and B12). The liver is composed of different types of cells, which mainly are liver sinusoidal endothelial cells, hepatocytes (present the largest constituent of the liver, 80–85%), hepatic stellate cells, and extracellular matrix (Fig. 4.12). The liver is built of four lobes, the large right lobe, left lobe, caudate lobe, and quadrate lobe, the last two lobes present between the right and left lobes. Microscopically, each liver lobe consists of hepatic lobules. The lobules are roughly hexagonal, consisting of sheets of hepatocytes radiating around a central vein. The central vein joins the hepatic vein to carry blood from the liver. The distinctive component of the lobule is the portal triad, which can be found at all corners of the lobule. The portal triad consists of five structures: a branch of the hepatic artery, a branch of the portal vein, a bile duct, a lymphatic vessel, and a branch of the vagus nerve [42]. The liver is connected by two large blood vessels: the hepatic artery, the portal vein, and the common hepatic duct. The hepatic artery carries oxygen-rich blood from the aorta through the celiac artery, while the portal vein carries nutrient-rich blood from the entire digestive system as well as from the spleen and pancreas (Fig. 4.12).

The precise configuration and variation of the liver cell types make liver tissue engineering using biomaterials difficult, and so fabrication of a tissue engineering scaffold similar to the liver nano- and mico-architecture is still a great challenge. Rad et al., prepared conducting scaffolds based on gelatin/chitosan with collagen, hyaluronan, or electroconductive poly(3,4-ethylenedioxythiophene) for liver tissue engineering. They tested the effect of scaffold composition on the proliferation of hepatocytes. Their results stated that the scaffolds contained conductive polymer together with collagen and hyaluronan enhanced attachment and proliferation of hepatocytes [45]. Ghahremanzadeh et al. synthesized a galactose-modified chitosan/ PCL electrospun scaffold for liver tissue engineering. This modification increased the hydrophilicity of the scaffolds. The results presented that treatment of chitosan surface with galactose enhanced obviously human hepatic (HepG2) cell growth and proliferation [46]. In a previous study, Wang et al., studied the effect of galactose moieties of galactosylated chitosan scaffold on hepatocyte cell attachment and growth. The results showed that galactose moieties improved hepatocyte cell attachment of the galactose [47]. Similarly, Shang et al. prepared a galactosylated chitosan/ hyaluronic acid scaffold freeze-drying technique to simulate the liver microenvironment. They designated the scaffold to be a favorable matrix for the co-culture of endothelial cells and hepatocytes. Their scaffold showed excellent bioactivity and viability against both types of cells, and it was recommended to be used as a novel co-culture model for liver drug discovery [48]. Growth and differentiation of hepatocyte cells are sensitive to extracellular matrix (ECM), and so the key to the success of liver tissue engineering is a selection of suitable ECM scaffold for hepatocytes. Li et al., prepared a polyethylene terephthalate fiber scaffold coated with ECM of decellularized human umbilical cord derived for liver tissue engineering application. Coating of scaffold fiber with ECM coating significantly enhanced HepaRG cell adhesion, proliferation, and differentiation [49].

Shortage of liver transplantation for patients suffering from complete loss of liver function has urged scientists to research and develop bioartificial liver systems based on hepatocytes seeded in a suitable scaffold and provide the bioreactor with the required nutrients and oxygen. Therefore, there have been numerous bioreactors and contained scaffolds developed for bioartificial liver models. For example, not limited to, Xia et al., established a cylindrical bioreactor as a bioartificial liver. It is composed of 12 double-sandwich culture plates, where, hepatocytes are placed between porous and nonporous collagen-coated polyethylene terephthalate membrane (Fig. 4.13).

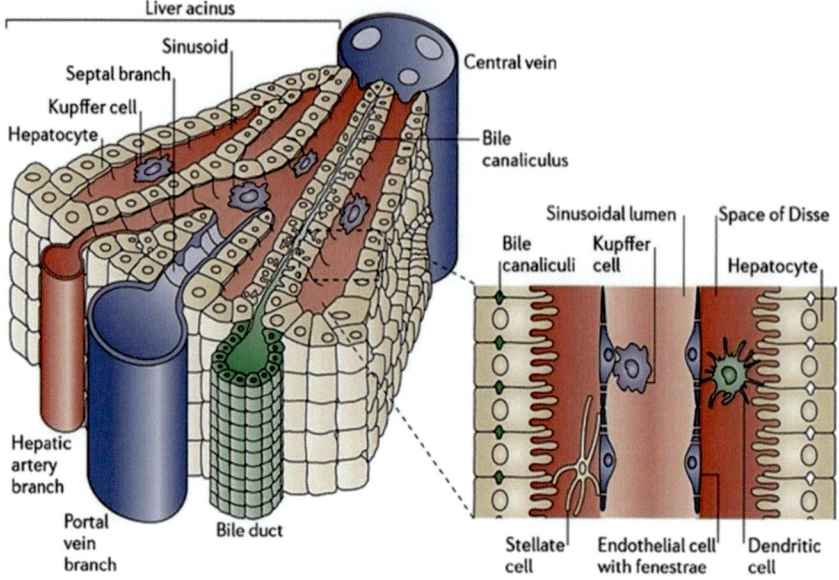

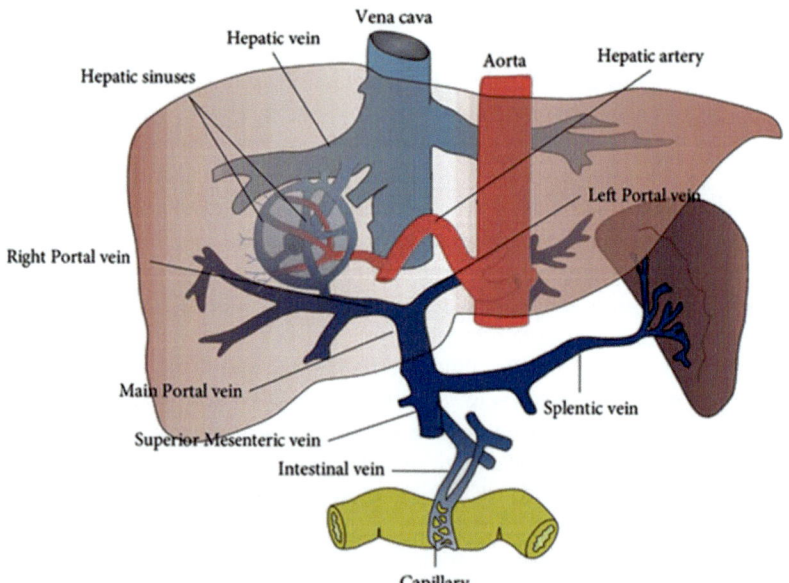

Fig. 4.12 Upper: portion of hepatic lobule: microanatomy [43], MDPI 2021, this article is an open access licensed under a Creative Commons Attribution 4.0 International License. Lower: hepatic blood flow [44], Willey Online Library 2022, this article is an open access licensed under a Creative Commons Attribution 4.0 International License

It possessed the ability to incubate up to 100 million hepatocytes which kept their phenotype and function [50]. Janani and Mandal fabricated a bioartificial liver model using liver ECM-functionalized silk scaffolds seeded with neonatal rat hepatocytes. As shown in Fig. 4.14, they simulate oxygen gradient by using periportal to the pericentral gradient in perfusion bioreactor contained scaffolds seeded with hepatocytes. Such bioreactor improved tissue remodeling, macrophage activation, decreasing of thrombogenicity and foreign-body response. Accordingly, the designed bioreactor was suggested to be useful in liver and drug screening [51]. Likewise, Li et al., fabricated a bioreactor with interconnected channels using PCL scaffold seeded with HepG2 cells followed by culturing with endothelial cells for 10 days. The results showed that HepG2 cells enhanced glucose metabolism, albumin production, and lactate production. As well as it allowing the perfusion of whole blood and solving the problem of blood compatibility limitation for liver tissue engineering [50]. Li et al., synthesized collagen-coated polyethylene terephthalate (PET) scaffold and studied its viability with hepatocytes (HepaRG cells). The results presented that a scaffold coated with collagen increased hydrophilicity and human liver progenitor HepaRG cells, albumin secretion, cell viability, and urea synthesis, and so they concluded that this scaffold can be applied to the bioartificial liver.

4.5 Vascular Tissue Regeneration

The human vascular system is a network of a large number of vessels that have a function of blood transformation in the circulatory system. There are three types of blood vessels; arteries, which transport blood from the heart to parts of the body, capillaries, which allow the exchange of water and chemicals between blood and tissues, and veins, which transport blood from the capillaries and return it to the heart. The word vascular—referring to blood vessels—is derived from the Latin word VAS, which means vessel. Arteries and veins are similar in that they have three layers, but the middle layer in arteries is thicker than that in veins (Fig. 4.15). The first one is endothelium (the thinnest layer) which is a single layer of simple squamous endothelial cells interconnected by a polysaccharide, surrounded by a thin layer of subendothelial connective tissue interlaced with a number of regular elastic bands called the lamina internal elastic. It rests on a connective tissue membrane with many elastic, collagenous fibers. This layer helps prevent blood clotting and relaxes the smooth muscle of the vessel by the release of nitric oxide. The middle layer (the thickest layer in arteries) is composed of elastic fibers of connective tissue and polysaccharides arranged circularly. It is separated from the next layer by a flexible layer called the outer elastic lamina. This layer controls the radius of the blood vessel. The outermost layer (the thickest layer in the veins), is composed entirely of connective tissue and contains the nerves that nourish the vessel and at the same time nourish the capillaries. Capillaries consist of an endothelial layer and sometimes connective tissue. However, blood vessels are not primarily involved in pumping blood as they lack peristaltic movement. But, arteries and some veins can

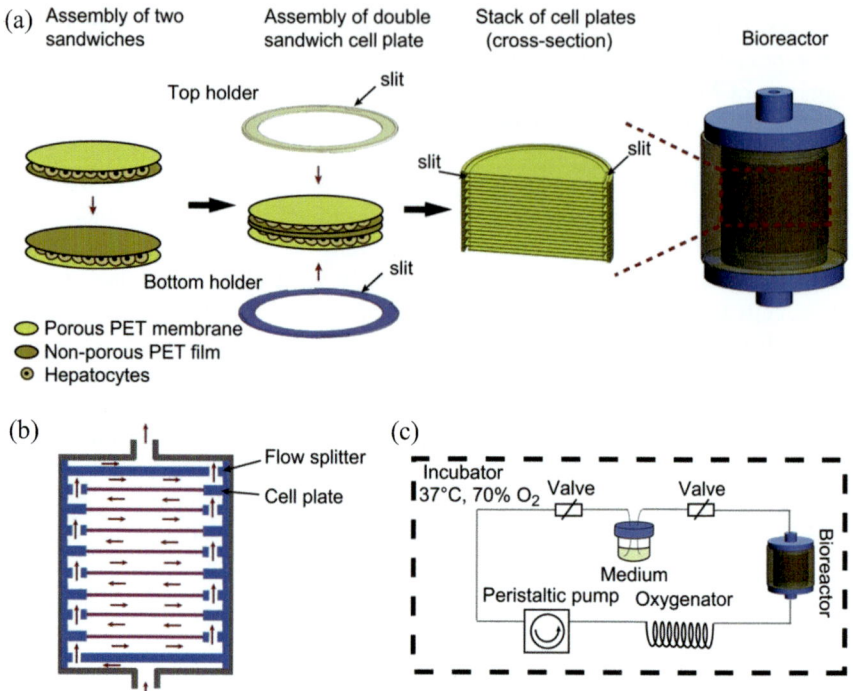

Fig. 4.13 Bioartificial liver using cylindrical bioreactor. **a** Interchange layer of hepatocytes and collagen-coated polyethylene terephthalate (PET) film with a sandwich-like structure. **b** Sandwich culture slices with inlet and outlet. **c** Bioreactor perfusion setup with a medium reservoir placed in the incubator at 37 °C and 70% O₂ [50]. Copyright Elsevier 2012

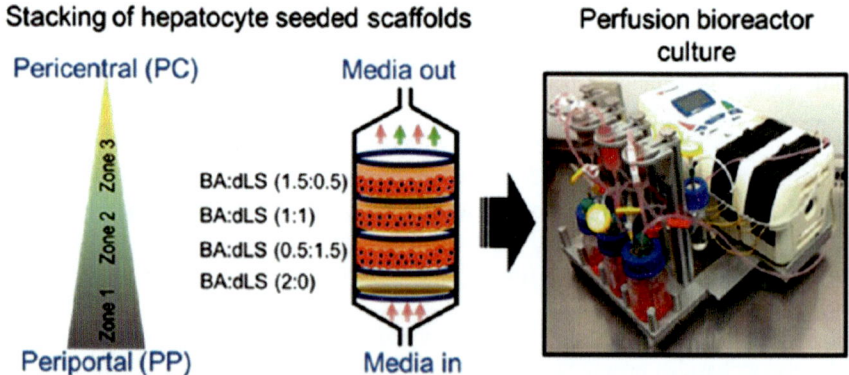

Fig. 4.14 Bioreactor ECM-functionalized silk scaffolds seeded with neonatal rat hepatocytes [51]. Copyright American Chemical Society, 2021

regulate their internal radius, by contracting the muscle layer surrounding them, this helps regulate the amount of blood coming to the organ, and this also contributes to the temperature regulation of the human body.

Vascular diseases, specifically atherosclerotic cardiovascular diseases, are one of the wide reasons for death [53]. It is caused usually by an accumulation of plaque on the interior side of the arterial wall leading to blood vessel narrowing and blockage. Arterial diseases can be treated by stent placement, bypass surgery, administration of anticoagulants, and vascular grafting. In specific and urgent cases, there is a need to replace the obstructed part of an artery which can be performed by autologous veins and arteries grafting taken from the same patient. Despite autologous grafts achieving gold-standard biocompatibility and the lowest risk of immune rejection, there is a patient percentage that, taken into consideration, does not possess suitable autologous grafts. Accordingly, the synthetic vascular graft can solve this shortage and unavailability; it is usually prepared by electrospinning technique. Vascular tissue engineering is considered a promising method to prepare synthetic vascular grafts. The success of vascular tissue engineering mainly depends on using of scaffold with a desirable porous structure, matched mechanical strength, and good cytocompatibility with vascular cells. Vascular tissue engineering can be performed by three approaches; in vitro, in vivo, and in situ (Fig. 4.16). In the case of in vitro vascular tissue engineering, the vascular scaffold is made by culturing the scaffold in a bioreactor including vascular cells. While in vivo vascular tissue engineering, the vascular scaffold is incubated inside the body (such as subcutaneous or peritoneal cavity) to give regenerated blood vessels. According to in situ vascular tissue engineering, synthetic cellular or acellular scaffold with required graft properties (e.g. porous structure, composition, degradability, and biocompatibility) can regenerate blood vessels in situ. In addition, the incorporation of anticoagulant drugs into vascular scaffolds is useful to prevent thrombosis. Nojavan et al., added aspirin (anticoagulant drug) into scaffold nanofiber based on carboxylated multi-walled carbon nanotubes and poly(ethylene terephthalate) as a vascular graft. They also studied the release profile and kinetics of aspirin and measured different physicochemical properties and hemocompatibility properties. The results showed that the addition of carbon nanotubes increased Young's modulus from 14.6 to 36.6 MPa and increased aspirin release up to 37%. Moreover, aspirin and carbon nanotubes decreased platelet adhesion and prolonged blood coagulation time. Finally, hemolysis results stated that all scaffolds can be in vivo used safely [54]. Yang et al., prepared a scaffold with vascular shape by electrospinning from silk fibroin and fibrin with different ratios. They also studied the degradation of different scaffolds in vivo subcutaneously in rat animal models (Fig. 4.17). Their findings demonstrated that silk fibroin increased mechanical properties but it decreased the scaffold hydrophilicity. In addition, all scaffolds possessed excellent viability with MSCs and blood. Moreover, the scaffolds of 75% fibrin in vivo degraded faster than those containing 100% silk fibroin [55].

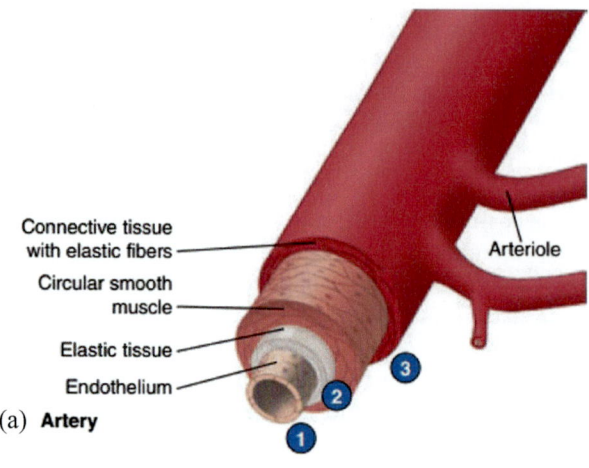

(a) **Artery**

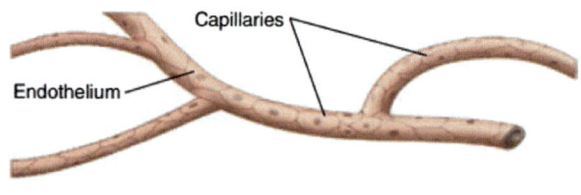

(b) **Capillary**

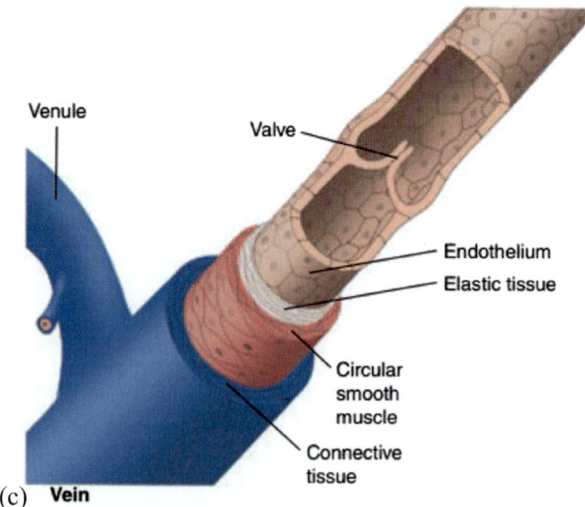

(c) **Vein**

Fig. 4.15 Vascular structure; **a** structure of artery, **b** structure of capillary, **c** structure of vein [52]

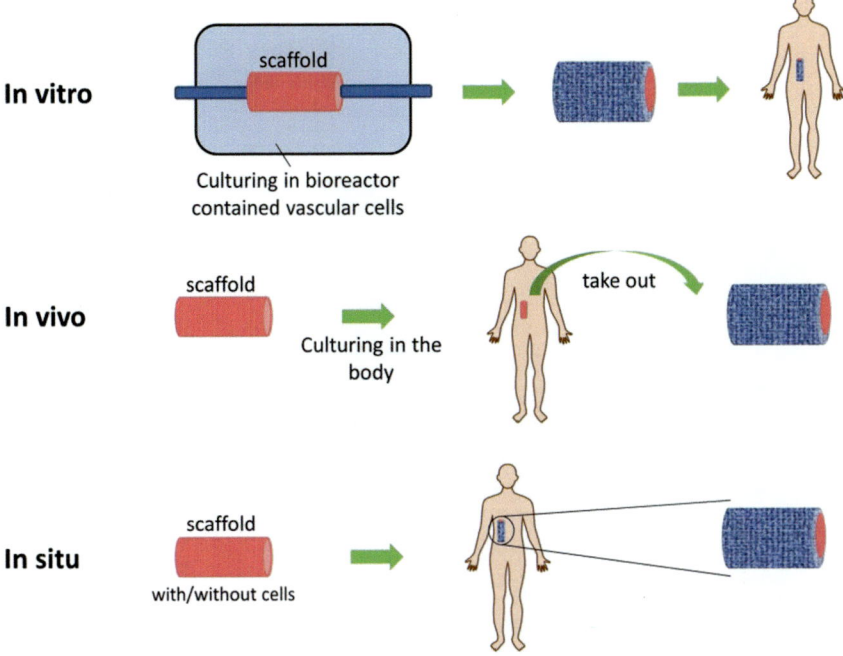

Fig. 4.16 Schematic presentation of different types of vascular tissue engineering; in vitro, in vivo, and in situ

4.6 Skeletal Muscle Tissue Regeneration

Muscle is a fibrous tissue characterized by the ability to contract and relax. It provides the movement of the organism. The structure of skeletal muscle is shown in Fig. 4.18. It consists of muscle bundles, and each bundle consists of muscle fibers. The muscle fiber is composed of muscle fibrils. A single fiber consists of adjacent muscle segments, and the muscle segments consist of protein filaments, namely actin and myosin. The cytoplasm of the muscle fiber is called sarcoplasm, and the membrane of the muscle fiber is called the sarcolemma. The muscle is divided into three sections: Striated skeletal muscle, which is a muscle consisting of a bundle of thin fibers, such as the muscles of the head, and limbs. It allows movement and is called voluntary muscle. Smooth muscle, is composed of rectangular cells or fibers and it is not connected to the skeleton, such as striated muscle, which surrounds hollow organs. Such as the intestines, trachea, and blood vessels, which are called involuntary muscles, and the heart muscle, which is also an involuntary muscle, but is closer in structure to skeletal muscle, and it is found only in the heart.

The skeletal muscle system is important for daily activity and movement, and any injury in it affects the lifestyle and activity, and it can cause permanent disability. Skeletal muscle injuries usually occur from traffic accidents, overload exercising,

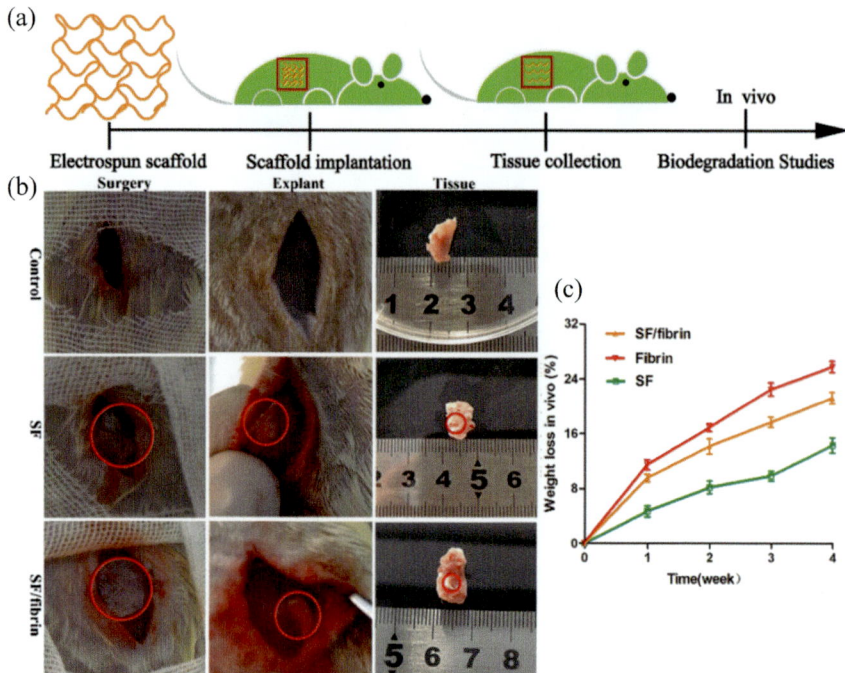

Fig. 4.17 Schematic presentation of in vivo subcutaneous degradation test (**a**). Subcutaneous implantation of the different vascular scaffolds (**b**). The weight loss curve of different scaffolds (**c**). Adapted from [55]. Nature 2024, this article is an open access licensed under a Creative Commons Attribution 4.0 International License

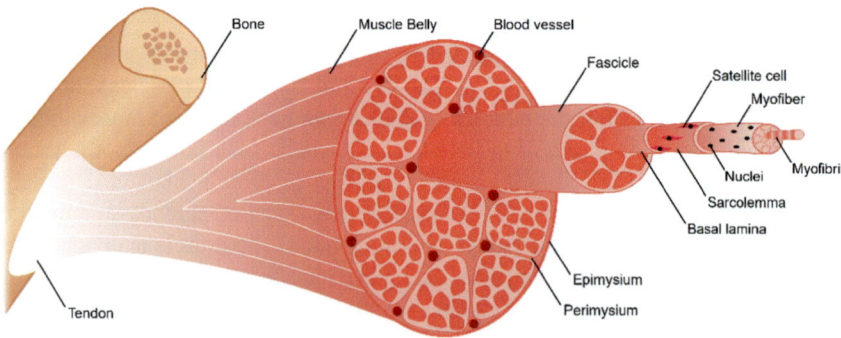

Fig. 4.18 Structure of skeletal muscle [61]. MDPI 2020, this article is an open access licensed under a Creative Commons Attribution 4.0 International License

contusions while doing a sport, and orthopedic surgeries. These injuries result in muscle tissue loss [56]. In case of small muscle loss caused by injury, skeletal muscle can regenerate spontaneously, but when the muscle tissue loss is more than 20%, it requires regenerative surgery. Muscle regeneration can be carried out by surgery, physical treatment, cell therapy, and muscular tissue engineering. Each method of muscle injury treatment can be useful for a specific case. Meanwhile, the muscular tissue engineering approach is suitable and effective for large muscle tissue loss. Herein, muscle tissue can be regenerated via stimulating new muscle formation by implanting a scaffold with a structure similar to muscle tissue. Alternatively, muscle tissue can be in vitro constructed by cells, scaffold, and molecular signaling in a bioreactor and subsequently implanted in the body [57].

Biomaterials have been widely used to fabricate scaffolds applied for muscle tissue regeneration. The success key of muscle tissue regeneration is guiding myoblasts to grow in an oriented direction by the scaffold. This requires fabricating an anisotropic scaffold and applying an external stimulator, such as current. Chen et al., prepared electroconductive and highly aligned and nanofibrous polyaniline/PCL scaffold for skeletal muscle regeneration application. They obtained aligned fibers formed by the application of an external magnetic field on the collector during the electrospinning process. Scaffolds with randomly oriented fibers were also prepared for comparison. The scaffold guidance ability was investigated using mouse C2C12 myoblasts. The results showed that a conductive and aligned scaffold promoted myotube formation and maturation, and it guided myoblast orientation [58]. Shah et al., synthesized phosphate glass fibers-reinforced collagen scaffolds for muscular tissue engineering. These composite constructs guided the proliferation of myocyte cells in one direction parallel to glass fibers, and gene expression studies demonstrated the formation and maturation of myofibre [59]. Quint et al., synthesized gelatin methacrylate/LAPONITE® nano clay composite hydrogel scaffold. Nanoclay was used to regulate and control the release of myogenic growth factor (insulin-like growth factor 1 (IGF-1)). The results exhibited that nano clay could control and retain 50% of the released IGF-1, and so it improved mouse myoblast C2C12 proliferation and differentiation [60].

References

1. Taichman RS (2005) Blood and bone: two tissues whose fates are intertwined to create the hematopoietic stem-cell niche. Blood 105:2631–2639
2. Tortora GJ, Nielsen M (2017) Principles of human anatomy. Wiley
3. Kovářík T, Křenek T, Bělský P, Šesták J (2017) Biomaterials and nanotechnology approach to medical enhancement. In: Šesták J, Hubík P, Mareš JJ (eds) Thermal physics and thermal analysis: from macro to micro, highlighting thermodynamics, kinetics and nanomaterials. Springer International Publishing, Cham, pp 449–470
4. Collins MN, Ren G, Young K, Pina S, Reis RL, Oliveira JM (2021) Scaffold fabrication technologies and structure/function properties in bone tissue engineering. Adv Func Mater 31:2010609

5. Breeland G, Sinkler MA, Menezes RG (2023) Embryology, bone ossification. StatPearls [Internet]. StatPearls Publishing
6. Lopes D, Martins-Cruz C, Oliveira MB, Mano JF (2018) Bone physiology as inspiration for tissue regenerative therapies. Biomaterials 185:240–275
7. Quarto R, Giannoni P (2016) Bone tissue engineering: past–present–future. In: Mesenchymal stem cells: methods and protocols, pp 21–33
8. Salamanca E, Hsu C-C, Huang H-M, Teng N-C, Lin C-T, Pan Y-H et al (2018) Bone regeneration using a porcine bone substitute collagen composite in vitro and in vivo. Sci Rep 8:984
9. Daculsi G, Fellah B, Miramond T, Durand M (2013) Osteoconduction, osteogenicity, osteoinduction, what are the fundamental properties for a smart bone substitutes. IRBM 34:346–348
10. Xia H, Dong L, Hao M, Wei Y, Duan J, Chen X et al (2021) Osteogenic property regulation of stem cells by a hydroxyapatite 3D-hybrid scaffold with cancellous bone structure. Front Chem 9:798299
11. Wu R, Li Y, Shen M, Yang X, Zhang L, Ke X et al (2021) Bone tissue regeneration: the role of finely tuned pore architecture of bioactive scaffolds before clinical translation. Bioact Mater 6:1242–1254
12. Lee J, Farag MM, Park EK, Lim J, Yun H (2014) A simultaneous process of 3D magnesium phosphate scaffold fabrication and bioactive substance loading for hard tissue regeneration. Mater Sci Eng C 36:252–260
13. Farag MM, Yun H (2014) Effect of gelatin addition on fabrication of magnesium phosphate-based scaffolds prepared by additive manufacturing system. Mater Lett 132:111–115
14. Raghunath J, Salacinski HJ, Sales KM, Butler PE, Seifalian AM (2005) Advancing cartilage tissue engineering: the application of stem cell technology. Curr Opin Biotechnol 16:503–509
15. Buckwalter J, Mankin H (1997) Articular cartilage: part I. J Bone Joint Surg 79:600
16. Li Y, Yuan Z, Yang H, Zhong H, Peng W, Xie R (2021) Recent advances in understanding the role of cartilage lubrication in osteoarthritis. Molecules 26:6122
17. Slattery C, Kweon CY (2018) Classifications in brief: outerbridge classification of chondral lesions. Clin Orthop Relat Res® 476:2101–2104
18. Wang K-Y, Jin X-Y, Ma Y-H, Cai W-J, Xiao W-Y, Li Z-W et al (2021) Injectable stress relaxation gelatin-based hydrogels with positive surface charge for adsorption of aggrecan and facile cartilage tissue regeneration. J Nanobiotechnol 19:1–16
19. Li W-J, Tuli R, Okafor C, Derfoul A, Danielson KG, Hall DJ et al (2005) A three-dimensional nanofibrous scaffold for cartilage tissue engineering using human mesenchymal stem cells. Biomaterials 26:599–609
20. Alves da Silva M, Martins A, Costa-Pinto A, Costa P, Faria S, Gomes M et al (2010) Cartilage tissue engineering using electrospun PCL nanofiber meshes and MSCs. Biomacromolecules 11:3228–3236
21. Heirani-Tabasi A, Hosseinzadeh S, Rabbani S, Tafti SHA, Jamshidi K, Soufizomorrod M et al (2021) Cartilage tissue engineering by co-transplantation of chondrocyte extracellular vesicles and mesenchymal stem cells, entrapped in chitosan–hyaluronic acid hydrogel. Biomed Mater 16:055003
22. Haghighi P, Shamloo A (2021) Fabrication of a novel 3D scaffold for cartilage tissue repair: in-vitro and in-vivo study. Mater Sci Eng C 128:112285
23. Kim YS, Majid M, Melchiorri AJ, Mikos AG (2019) Applications of decellularized extracellular matrix in bone and cartilage tissue engineering. Bioeng Transl Med 4:83–95
24. Stone RN, Frahs SM, Hardy MJ, Fujimoto A, Pu X, Keller-Peck C et al (2021) Decellular-ized porcine cartilage scaffold; validation of decellularization and evaluation of biomarkers of chondrogenesis. Int J Mol Sci 22:6241
25. Li Y, Xun X, Xu Y, Zhan A, Gao E, Yu F et al (2022) Hierarchical porous bacterial cellu-lose scaffolds with natural biomimetic nanofibrous structure and a cartilage tissue-specific microenvironment for cartilage regeneration and repair. Carbohyd Polym 276:118790
26. Ayariga JA, Huang H, Dean D (2022) Decellularized avian cartilage, a promising alternative for human cartilage tissue regeneration. Materials 15:1974

27. MacKay D, Miller AL (2003) Nutritional support for wound healing. Altern Med Rev 8
28. Rhett JM, Ghatnekar GS, Palatinus JA, O'Quinn M, Yost MJ, Gourdie RG (2008) Novel therapies for scar reduction and regenerative healing of skin wounds. Trends Biotechnol 26:173–180
29. Dermatology K. https://konadermatologistcom/anatomy-of-the-skin/
30. Trinh X-T, Long N-V, Van Anh LT, Nga PT, Giang NN, Chien PN et al (2022) A comprehensive review of natural compounds for wound healing: targeting bioactivity perspective. Int J Mol Sci 23:9573
31. Sarkar SD, Farrugia BL, Dargaville TR, Dhara S (2013) Chitosan–collagen scaffolds with nano/microfibrous architecture for skin tissue engineering. J Biomed Mater Res Part A Off J Soc Biomater Jpn Soc Biomater Aust Soc Biomater Korean Soc Biomater 101:3482–3492
32. Khalilimofrad Z, Baharifar H, Asefnejad A, Khoshnevisan K (2023) Collagen type I cross-linked to gelatin/chitosan electrospun mats: application for skin tissue engineering. Mater Today Commun 35:105889
33. Sun L, Li J, Gao W, Shi M, Tang F, Fu X et al (2021) Coaxial nanofibrous scaffolds mimicking the extracellular matrix transition in the wound healing process promoting skin regeneration through enhancing immunomodulation. J Mater Chem B 9:1395–1405
34. Xu D, Feng Y, Song M, Zhong X, Li J, Zhu Z et al (2024) Smart and bioactive electrospun dressing for accelerating wound healing. Chem Eng J 496:153748
35. Pazhouhnia Z, Noori A, Farzin A, Khoshmaram K, Hoseinpour M, Ai J et al (2024) 3D-Bioprinted GelMA/gelatin/amniotic membrane extract (AME) scaffold loaded with keratinocytes, fibroblasts, and endothelial cells for skin tissue engineering. Sci Rep 14:12670
36. Girard F, Lajoye C, Camman M, Tissot N, Berthelot Pedurand F, Tandon B et al (n/a) First advanced bilayer scaffolds for tailored skin tissue engineering produced via electrospinning and melt electrowriting. Adv Funct Mater 2314757
37. Isseroff RR, Dahle SE (2012) Electrical stimulation therapy and wound healing: where are we now? Adv Wound Care 1:238–243
38. Liu W, Zhu Y, Tao Z, Chen Y, Zhang L, Dong A (2023) Black phosphorus-based conductive hydrogels assisted by electrical stimulus for skin tissue engineering. Adv Healthc Mater 12:2301817
39. Aycan D, Selmi B, Kelel E, Yildirim T, Alemdar N (2019) Conductive polymeric film loaded with ibuprofen as a wound dressing material. Eur Polym J 121:109308
40. Najafian S, Eskandani M, Derakhshankhah H, Jaymand M, Massoumi B (2023) Extracellular matrix-mimetic electrically conductive nanofibrous scaffolds based on polyaniline-grafted tragacanth gum and poly(vinyl alcohol) for skin tissue engineering application. Int J Biol Macromol 249:126041
41. Ud-Din S, Bayat A (2014) Electrical stimulation and cutaneous wound healing: a review of clinical evidence. Healthcare 2:445–467
42. Simpson ML (2014) Human anatomy & physiology
43. Panwar A, Das P, Tan LP (2021) 3D hepatic organoid-based advancements in LIVER tissue engineering. Bioengineering 8:185
44. Key A. https://aneskeycom/hepatic-physiology-anesthesia/
45. Tahmasbi Rad A, Ali N, Kotturi HSR, Yazdimamaghani M, Smay J, Vashaee D et al (2014) Conducting scaffolds for liver tissue engineering. J Biomed Mater Res Part A 102:4169–4181
46. Ghahremanzadeh F, Alihosseini F, Semnani D (2021) Investigation and comparison of new galactosylation methods on PCL/chitosan scaffolds for enhanced liver tissue engineering. Int J Biol Macromol 174:278–288
47. Wang B, Hu Q, Wan T, Yang F, Cui L, Hu S et al (2016) Porous lactose-modified chitosan scaffold for liver tissue engineering: influence of galactose moieties on cell attachment and mechanical stability. Int J Polym Sci 2016:2862738
48. Shang Y, Tamai M, Ishii R, Nagaoka N, Yoshida Y, Ogasawara M et al (2014) Hybrid sponge comprised of galactosylated chitosan and hyaluronic acid mediates the co-culture of hepatocytes and endothelial cells. J Biosci Bioeng 117:99–106

49. Li Y, Zhang Y, Zhong K, Liao S, Zhang G (2024) The development of a 3D PET fibrous scaffold modified with an umbilical cord dECM for liver tissue engineering. Polymers 16:1794
50. Xia L, Arooz T, Zhang S, Tuo X, Xiao G, Susanto TAK et al (2012) Hepatocyte function within a stacked double sandwich culture plate cylindrical bioreactor for bioartificial liver system. Biomaterials 33:7925–7932
51. Janani G, Mandal BB (2021) Mimicking physiologically relevant hepatocyte zonation using immunomodulatory silk liver extracellular matrix scaffolds toward a bioartificial liver platform. ACS Appl Mater Interfaces 13:24401–24421
52. Pharmacy180. https://www.pharmacy180.com/article/blood-vessel-structure-3624/
53. Mortality G (2016) Causes of death collaborators global, regional, and national life expectancy, all-cause and cause-specific mortality for 249 causes of death, 1980–2015: a systematic analysis for the global burden of disease study 2015. Lancet 388:1459–1544
54. Nojavan C, Sepehri R, Harirchi P, Zahedi P, Kabiri M, Kharat Z et al (2024) Potential use of electrospun poly(ethylene terephthalate)/Carbon nanotubes containing aspirin in vascular tissue engineering application. Fibers Polym 25:71–81
55. Yang L, Wang X, Xiong M, Liu X, Luo S, Luo J et al (2024) Electrospun silk fibroin/fibrin vascular scaffold with superior mechanical properties and biocompatibility for applications in tissue engineering. Sci Rep 14:3942
56. Pollot BE, Corona BT (2016) Volumetric muscle loss. In: Skeletal muscle regeneration in the mouse: methods and protocols, pp 19–31
57. Wang L, Cao L, Shansky J, Wang Z, Mooney D, Vandenburgh H (2014) Minimally invasive approach to the repair of injured skeletal muscle with a shape-memory scaffold. Mol Ther 22:1441–1449
58. Chen M-C, Sun Y-C, Chen Y-H (2013) Electrically conductive nanofibers with highly oriented structures and their potential application in skeletal muscle tissue engineering. Acta Biomater 9:5562–5572
59. Shah R, Knowles JC, Hunt NP, Lewis MP (2016) Development of a novel smart scaffold for human skeletal muscle regeneration. J Tissue Eng Regen Med 10:162–171
60. Quint JP, Samandari M, Abbasi L, Mollocana E, Rinoldi C, Mostafavi A et al (2022) Nano-engineered myogenic scaffolds for skeletal muscle tissue engineering. Nanoscale 14:797–814
61. Carnes ME, Pins GD (2020) Skeletal muscle tissue engineering: biomaterials-based strategies for the treatment of volumetric muscle loss. Bioengineering 7:85

Chapter 5
Recent Trends and Challenges of Biomaterials for Tissue Regeneration

Abstract This chapter shows new trends and perspectives on tissue regeneration. Involving gene therapy in tissue regeneration is considered a promising method to enhance tissue regeneration with minimum immunological response. Artificial intelligence is a new way applied recently in regenerative medicine using big data to reduce the time and cost of experimental tests of new biomaterials and drugs. The chapter refers to potential new trends, such as the development of new biomaterials, 3D bioprinting technique, and using of stem cells, and consequently genetically modified stem cells in the tissue engineering field. Finally, the authors present their view about the possibility of fabricating fully functional organs in the future.

5.1 Genetic Application in Tissue Regeneration

Recently, gene therapy has been considered a hopeful approach for the treatment of incurable diseases, and scientists have great hopes for its use in the achievement of promising results in the tissue regeneration field. Gene therapy is defined according to NIH as "a method that uses a gene(s) to treat, prevent or cure a disease or medical disorder". It is defined according to FDA as "a treatment route to modify or manipulate the expression of a gene or to alter the biological properties of living cells for therapeutic use". In this technique, healthy new gene copies are added to or replace a defective or missing gene in a patient's cells. It is well-known that each human cell contains 23 pairs of chromosomes present in the nucleus, each chromosome composed of DNA (deoxyribonucleic acid). DNA is made of two twisted strands with the shape of a spiral stair called a helix. It is composed of small four chemical bases called nucleotides; adenine (A), guanine (G), cytosine (C), and thymine (T) (Fig. 5.1). Gene is a short section of DNA, it tells the cell to make proteins that carry out specific tasks depending on cell type. For example, hair and fingernails are made of linked keratin proteins. Such protein is formed by keratinocytes in the middle of the epithelium via the order of specific keratin genes that are responsible for the formation of these types of cells.

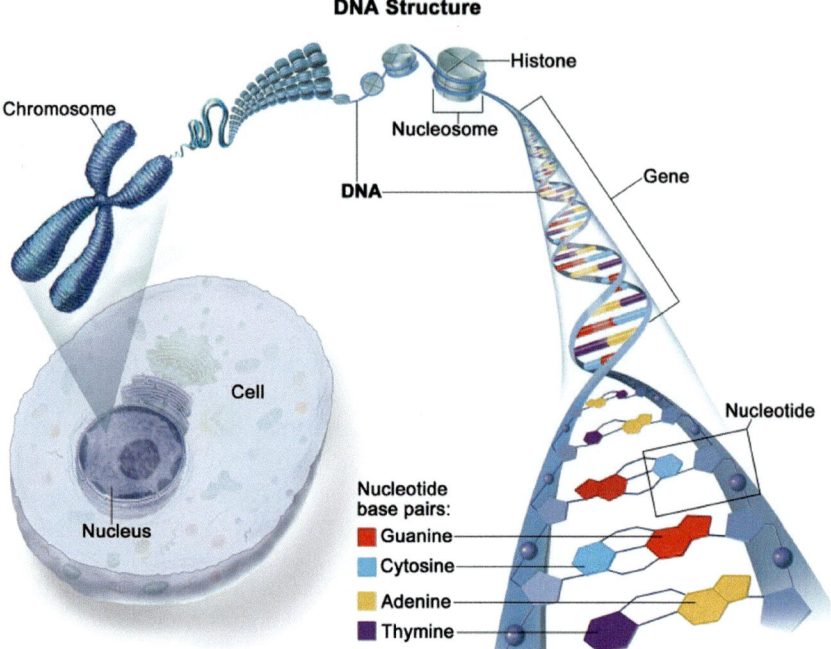

Fig. 5.1 Structure of a gene is a section of DNA chain that forms a chromosome. DNA is composed of two strands that twist into a helix stair shape. DNA is made up of four building bases called nucleotides: adenine (A), thymine (T), guanine (G), and cytosine (C). By National Cancer Institute (NCI) [1]. Free copyright

The development of regenerative medicine depends on understanding the mechanism of formation and regeneration of the tissue. Genes contribute to tissue reconstruction and repair. In this respect, the discovery of gene roles helps the rapid development of gene therapy, which is becoming a promising approach for tissue regeneration [2, 3]. Table 5.1 shows examples of therapeutic gene classes applied in gene therapy. There are two methods of gene therapy; direct (in vivo) gene therapy to obtain gene vectors or vehicles, and indirect (ex vivo) gene therapy to get genetically modified cells (Fig. 5.2). In both ways, the gene can be transferred by either viral vectors (transduction) or nonviral vectors (transfection). Viral vector uses the virus as a gene carrier. This method is characterized by high transfer efficiency and large gene capacity. However, the problems of this method are toxicity, unstable episomes, immunogenicity, and short-term gene expression. Nonviral vectors do not utilize the virus as a gene carrier. The advantages of this method are low cost, simple, non-toxic, no immunogenicity, and large gene capacity. But, its limitations are a high rate of degradation, low transfer efficiency, short-term gene expression, and dividing cells [4].

Table 5.1 Classes of genes with examples used in tissue regeneration

Class	Example genes	Abbreviation	Function
Cytokines	Interleukin 4 protein	IL-4	They are signaling proteins that contribute in inflammation control
	Brain-derived neurotrophic factor	BDNF	
	Bone morphogenetic protein-2	BMP-2	
	Erythropoietin protein	EPO	
	Transforming growth factor-beta 1	TGF-β1	
	Growth differentiation factor 5	GDF5	
Cellular functional proteins	Telomerase reverse transcriptase	hTERT	They carry out most of cell functions
	Fibroblast growth factor 7	FGF-7	
	NK2 homeobox 1	NKX2-1	
	Paired box gene 8	PAX8	
	Dyskerin pseudouridine synthase 1	DKC1	
	Liver-enriched transcription factors	LETFs	
	GATA binding protein 6	GATA6	
	Leucine rich repeat kinase 2	LRRK2	
Transmembrane proteins	Polycystin-1 Polycystin-2	PKD-1 PKD-2	Transportation of ions and molecules, cell signaling, and participate in the structural integrity of the cell membrane
	Cystic fibrosis transmembrane conductance regulator	CFTR	

(continued)

Table 5.1 (continued)

Class	Example genes	Abbreviation	Function
Cytoskeleton proteins	Intraflagellar transport protein 140	IFT140	Assist cells maintain their shape and internal organization and provide mechanical support
Fluorescence proteins	Green fluorescent protein mCherry	GFP mCherry	Track concentrations of small-molecule metabolites, protein conformational changes, and enzyme activities in living cells

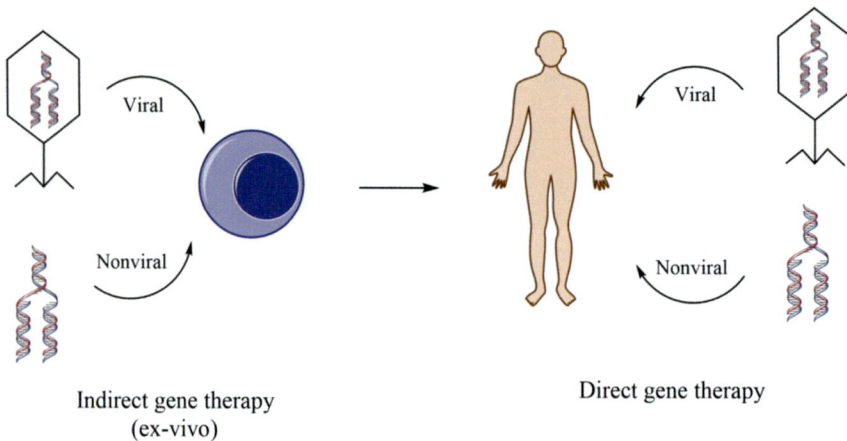

Fig. 5.2 Schematic presentation of direct and indirect (ex-vivo, by genetically modified cell) gene therapy by viral and nonviral methods

A genetically modified cell is considered a distinctive technique used to deliver genes in tissue engineering. It can induce cell differentiation and produce the necessary proteins required for successful tissue regeneration [5]. Genetically modified cells can be applied to classical two-dimensional monolayer cells or cell spheroids. Cell spheroids are supreme carriers for gene delivery tissue regeneration. Cell spheroids are three-dimensional cell aggregates cell cultures that self-aggregate into sphere-like formations during cell proliferation, and they mimic diseased and normal tissues (Fig. 5.3). Cell spheroids have an in vivo survival rate higher than non-spheroid cells [6]. Moreover, they have been used for drug screening, construction of disease models, tumor treatment by sending suicide signals to the cancer cells, and stimulation of bone and cartilage regeneration.

Cell spheroids can be obtained by scaffold-based or scaffold-free methods. In the scaffold-based method, hydrogels (e.g. hyaluronic acid, alginate, agarose, and collagen), biofilms (e.g. bacterial cellulose), and particles (e.g. magnetic nanoparticles, nanofiber particles, and polymer microspheres) are used as solid scaffolds

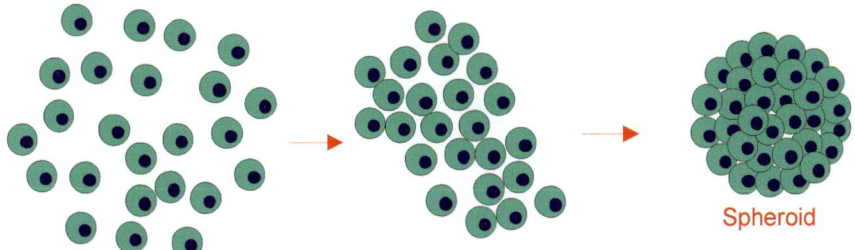

Fig. 5.3 Schematic presentation of cell spheroid formation. Adapted from [8]. MDPI 2017, this article is an open access licensed under a Creative Commons Attribution 4.0 International License

that provide microenvironment for cell aggregation. This method is easy and cost-effective; however, it has limitations, such as scaffold instability and difficulty in monitoring spheroids due to interference with scaffold material [7]. The scaffold-free method is recently more desirable than the scaffold-based method because of avoids scaffold instability and interference. In this method, the cells are aggregated physically, like spinning culture by continuous stirring of cell suspension, pellet cell culture in non-adhesive bottom palates, and magnetic culturing using magnetic hydrogel particles which are subjected to magnetic force to agglomerate the cells.

Biomaterial-guided gene delivery is a safe and promising nonviral gene method to transfer genes into the cell. Carballo-Pedrares et al., synthesized gene-activated hyaluronic acid cryogels for cartilage tissue regeneration. They synthesized nonviral vectors of plasmids encoding for β-galactosidase (placZ) and SOX9 based on niosomes (P80PX nioplexes or LPF lipoplexes), and they introduced them in hyaluronic acid cryogel (HACG) functionalized with methacrylate groups. Finally, the cyrogel was seeded with human mesenchymal stem cells (hMSCs), and they conducted chondrogenesis tests in situ and an ex-vivo model of the chondral defect. The results showed that integration of a plasmid encoding for the transcription factor SOX9 into HACG-P80PX systems resulted in a clear MSCs chondrogenic differentiation. Moreover, the test of ex-vivo model of chondral defect demonstrated the ability of this system to restore cartilage extracellular matrix [9].

Li et al., incorporated calcitonin gene-related peptide in 13-93B1.5 borosilicate bioactive glass scaffold for tissue regeneration. They prepared glass based on 6-Na_2O–8 K_2O–8 MgO–22 CaO–27 SiO_2–27B_2O_3–2P_2O_5 mol.% by melting method and grounded it to a particle size range of 20–40 μm. The glass powder was used to fabricate scaffolds with robocasting technology. The final scaffolds were functionalized with sodium alginate/calcitonin gene-related peptide (CGRP) hydrogel. The osteogenesis capacity of functionalized scaffolds was evaluated against Human bone marrow mesenchymal stem cells (HBMSCs). As well as in vivo subcutaneous implantation test was carried out. The results showed that CGRP-functionalized scaffold promoted osteogenic differentiation [10].

Malek-Khatabi et al., used PCL nanofibrous scaffold functionalized with metalloprotease-sensitive peptide to immobilize plasmid DNA (pDNA)-based/

chitosan nanocomplex for bone tissue regeneration. pDNA encoding BMP-2 which is an important gene for bone regeneration. Osteogenesis tests were performed on bone marrow-derived mesenchymal stem cells (BMMSCs). The results of the nanocomplex/scaffolds gene delivery system improved osteogenesis, and it is expected to be applied successfully in bone regenerative medicine.

Carvalho et al., designed local gene delivery of plasmid DNA encoding for enhanced green fluorescent protein (pEGFP) embedded in lipoplexes. They used alginate-based microgels composed of chondroitin sulfate, poly(allylamine hydrochloride), and poly-l-lysine fabricated by droplet-based microfluidics. These polyelectrolyte microgels showed low cytotoxicity and the microgel-loaded gene demonstrated the ability for surface-based transfection of MCF-7 cells [11].

Lee et al., immobilized platelet-derived growth factor (PDGF) by fragmented PLC fibers coated with polydopamine and subsequently coated with gelatin and biomineralized their surfaces by Ca-phosphate biomineral by immersion in SBF, thereafter. The fibers were used to prepare human adipose-derived stem cells (hADSCs) spheroids for vascularized bone regeneration (Fig. 5.4). The results showed that incorporation of PDGF in the spheroid improved the proliferation, enhanced osteogenic differentiation, and enhanced endothelial differentiation of hADSCs. Moreover, the transplantation results of genetically modified spheroids in vivo mouse calvarial defect presented improved formation of new bone and number of capillaries. Accordingly, these genetically modified together with biomineralized fiber were strongly suggested for vascularized bone regeneration [12].

5.2   Artificial Intelligence in Tissue Regeneration

Tissue engineering still needs more development and improvement. For example, biomaterials used up to date in tissue regenerative medicine did not achieve the ideal properties starting from architecture and ending to lack of vascularization ability and necessary biomolecules. Therefore, each material construct has to be evaluated and tested manually before it is applied clinically in the regeneration of diseased or lost tissue which requires time, manpower, and costs. These obstacles can be tackled by artificial intelligence (AI). AI can simulate human complicated mental processes. This can happen when the machines are fed with enough information and entries, and in this case, the machines will have human-like abilities in logical thinking and perception [13]. And so, they are considered a powerful aid in predicting tissue response against the material constructs used in tissue regeneration. Generally, AI contributes extensively to the healthcare sector; it is utilized to assist in diagnostics and the determination of suitable disease treatment methods. Specifically, AI develops the tissue engineering field through biomaterial geometry design, cell culture optimization and monitoring, computational modeling, gene therapy, design of precise geometry of 3D scaffolds, image analysis of medical imaging, and controlling for the microfabrication process of organoids [14].

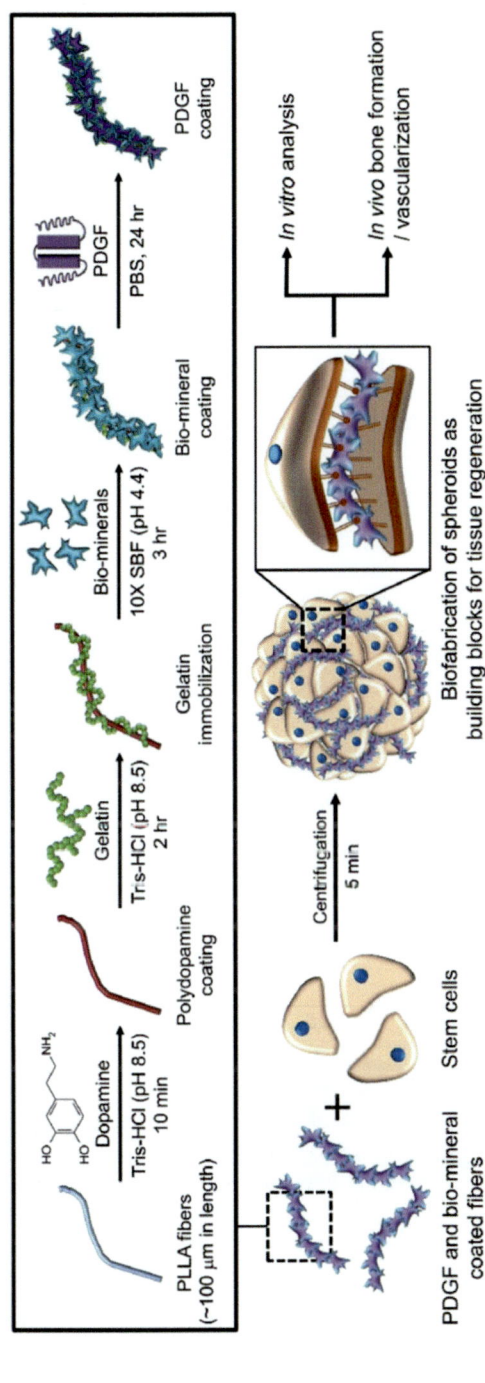

Fig. 5.4 Schematic presentation of preparation of human adipose-derived stem cells (hADSCs) spheroids modified by platelet-derived growth factor (PDGF) [12]. Copyright Elsevier 2020

AI includes subfields such as machine learning (ML) and deep learning (DL). ML is a process of continuous improvement of the accuracy of data analysis made by a machine via unique algorithms. DL is a subset of ML typically applied throughout multi-layer neural networks [15]. AI uses conventional artificial neural networks to solve complicated equations based on entries to make a decision. Artificial neural networks are similar to biological neural networks. Where both build of nodes (neurons) linked by synapses (Fig. 5.5). In the biological neural networks, the impulse collected by dendrites moves toward the nervous cell nucleus, and then the impulses are carried away from the nucleus body towards axon terminals through the axon, and this output signal becomes an input signal to another neuron. Similarly, in artificial neural networks, each node sums the data output of previous nodes and processes these inputs in an activation function to give an output that becomes new inputs to another node [16].

AI can be applied in drug discovery, clinical imaging, genomics, biomedical devices, and healthcare. Most drugs and biomedical devices are examined on cells

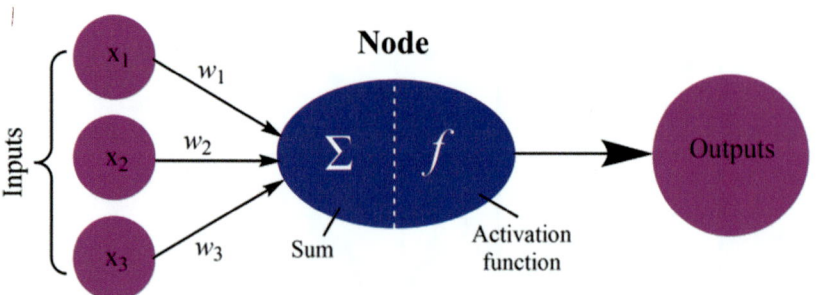

Fig. 5.5 Comparison between biological neural networks and artificial neural networks. Adapted from [16]. IOP Science 2021. This is an open access article

and animals, but the experimental results are not accurate because of the physiological differences between animals and humans [17]. Therefore, an interdisciplinary technology has developed to fabricate simulated microenvironments of human organs by a combination of cell biology, biomaterials, engineering, chemistry, and microfabrication on microfluidic chips, this is known as organ-on-a-chips (OOC) [18]. Marsano et al., fabricated heart-on-a-chip [19]. AI was used to analyze images (obtained by CT, micro-CT, MRI … etc.) of OOC. Kong et al. used ML approaches to predict the anticancer drug response on 3D fabricated organoid culture models (colorectal and bladder models) to predict the efficacy of anti-cancer drugs [20]. Figure 5.6 shows an example of OOC.

AI has been used in the prediction of an acceptability degree of certain materials to be applied in tissue regeneration. Prediction depends on the judgement of biocompatibility, angiogenesis, toxicity, biodegradability, and mechanical properties data from previous works. However, manual collection of big data is so difficult and time-consuming. Using AI to collect enables to extract data in an accurate, automated, and quick way from the text of available literature. Fuenteslopez et al., compared different mining-text tools for polydioxanone, because the research works concerned it is comparatively low, to predict its biocompatibility. Their results showed that each mining-text tool gave results different from another tool used in this work. This is attributed to multidisciplinary language and limited lexical resources of biomaterials [21]. Lu et al., applied AI (artificial neural network model) to estimate the antibacterial and cytocompatibility of titanium alloy functionalized with silver nanoparticles

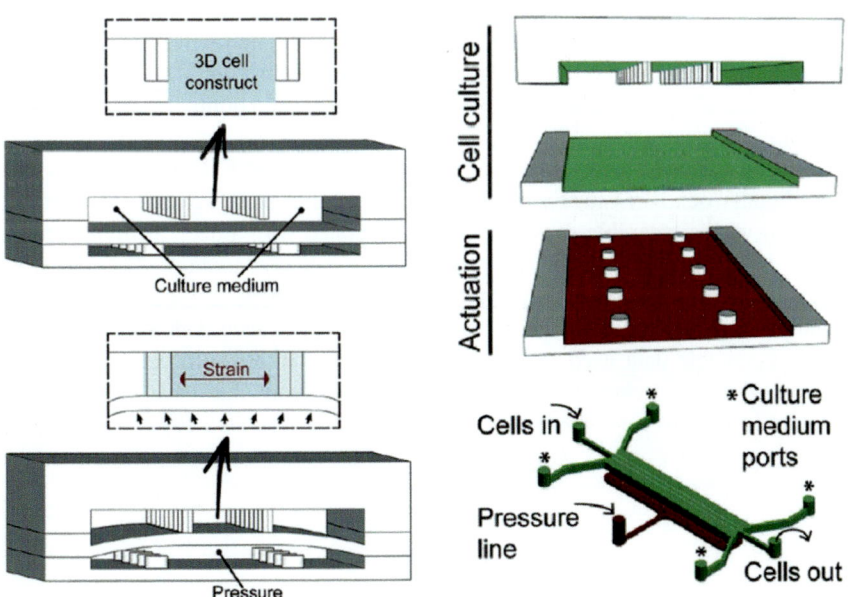

Fig. 5.6 Example of organ-on-a-chip platform. Adapted from [24]. Elsevier, this is an open access article

and its surface treated with the femtosecond laser to induce micro/nano surface structures. The artificial neural network model used in this study successfully predicted the relationship between micro/nano-structures and the femtosecond laser processing parameters to tailor the alloy properties [22]. Batoni et al., showed a computational model of hydrolytic degradation PLLA scaffold factionalized by bioactive protein (human ICOS-Fc) release of this protein for bone regeneration. Their study was based on a scaffold without macroporosity and a scaffold with internal macroporous structure with open. The results showed that the applied model can be used to improve the scaffold architecture and geometry to control scaffold degradation and protein release [23].

Barrera et al. [25] applied 3D CNNs (convolutional neural network) in medicine on digital slices of CAD models instead of real CT images for the prediction of the mechanical properties of tissue engineering scaffolds with different architectures (Fig. 5.7). The results showed that it could be possible to measure the mechanical strengths of scaffolds with different cell units by AI. Chen et al., utilized ML method to identify the morphology of human bone marrow stromal cells in poly(ε-caprolactone) (PCL) fibrous substrates prepared by electrospinning technique compared to flat films of the same polymer prepared by spin-coating approach. They used previous image databases and confocal microscope photos obtained in their study using an active contour model executed by a modified MATLAB script. Their results demonstrated that early cell response to fibrous substrate can be identified from the solidity, axis length, and concaving of cells [26]. Similarly, Tourlomousis et al., used ML metrology of confocal fluorescence microscope images with an automated single-cell bioimage data analysis to study the effect of 3D porous substrate architecture on the cell shape. They fabricated woven and non-woven PCL by electrospinning method. The combined data helped ML to classify cell shape phenotypes, and the results showed that woven substrates gave more of a homogenous cell shape than non-woven substrates [27].

AI can be used to monitor conditions of tissue regeneration. Kalasin et al., designed an intelligent wound dressing based on poly(vinyl acrylic) gel@PANI/Cu_2O hydrogel connected with a wearable sensor using AI (operated with a deep artificial neural network algorithm) to monitor and predict skin tissue regeneration. The sensor is based on the pH-responsive voltage output concept, and so the current resulting from wound pH change during the wound healing process was recorded. So, it provided necessary suitable data used by computer algorithms to assist in deciding on the selection of the best choice of wound dressing and drugs that gave the optimum treatment conditions for chronic wounds [28].

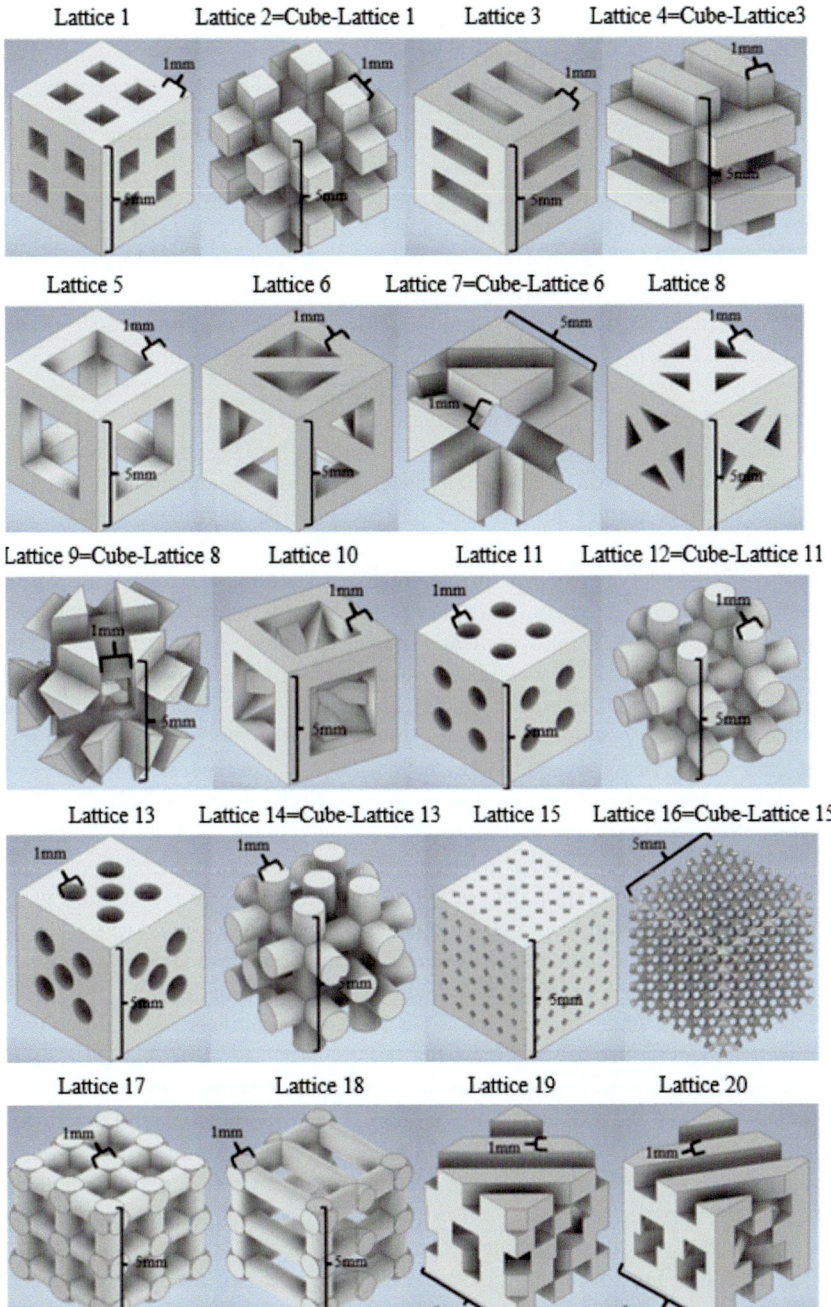

Fig. 5.7 Different tissue engineering scaffolds architecture designed by CAD library. Cell units of $5 \times 5 \times 5$ mm^3 are shown [25]. MDPI 2021, this article is an open access licensed under a Creative Commons Attribution 4.0 International License

5.3 Recent Trends and Challenges of Biomaterials for Tissue Regeneration

In previous chapters of this book, we demonstrated common applications of biomaterials in regeneration medicine for bone, skin, muscles, cartilage, liver, and vascular tissues. Despite success in regenerating diseased tissues and replacing damaged tissues, biomaterials still have shortages and limitations in the field of tissue regeneration from the side of material composition and architecture. According to composition, biomaterials must be nontoxic, biocompatible, forming bonds with the host tissue, stimulate angiogenesis, non-immunogenic, their degradation products do not cause harmful effects on other organs and can be modified by functional moiety. Concerning architecture, biomaterials are preferred to have porous structures with interconnected channels to achieve cell migration and get rid of cell wastes, as well as, their mechanical properties should be matched with the host tissue and do not alter as a function of time. For example, biomaterials used in bone regeneration should have an architecture that mimics the bone structure and provides angiogenesis and vascularization to deliver various important nutrients for osteogenesis activity, and zonal variation of cell density and bone structure which is an additional critical issue. Moreover, the mechanical strength of biomaterials used for bone regeneration should be as high as to support bone tissue, specifically in load-bearing parts. Cartilage regeneration is also still a critical issue that is because of the unique composition and structure of cartilage tissue. It is composed of collagen, mainly Type II, proteoglycans, and chondrocytes, and is subdivided into three zones with different structures and properties. And so, it is being argued to fabricate cartilage engineering scaffold materials with composition and architecture that mimics cartilage tissue, specifically, chondrocytes which are limited in availability [29], they synthesize all the extracellular matrix (ECM) components and regulate matrix metabolism, are highly specialized cells sparingly spread within ECM. In addition, cartilage tissue is avascular tissue which increases the struggle to regenerate it.

However, till now no prepared biomaterial encountered these ideal requirements. In order to defeat these limitations, there was a combination of different disciplines, such as material science, physics, biology, chemistry, engineering, and medicine have worked together to overcome these limitations. Wherever development of tissue regeneration biomaterials is going in two directions; the first one is the improvement of composition, surface roughness, morphology, gene modification, and functionalization. The second one is the elaboration of preparation methods and technologies to obtain materials with mimic architecture. Biomaterials are a very important candidate in the process of tissue regeneration. Besides the composition, treatment of biomaterial by physical and chemical methods potentially improves their biocompatibility and functionality. For instance, micro- and nano-pattering of biomaterial surfaces by laser or plasma methods were applied to create tailored roughness and improve surface hydrophilicity. Furthermore, biomaterials can possess antibacterial activity when they are functionalized with antibacterial agents (antibacterial ions, drugs, etc.).

However, obtaining excellent biocompatible material able to promote tissue regeneration, provide a suitable microenvironment for cell migration and proliferation, and stimulate angiogenesis remains a challenge. Therefore, the inclusion of nanotechnology, bioactive molecules (such as proteins, growth factors, and nucleic acids), gene therapy, and cells (specifically, cell spheroids) are recent new trends in tissue regeneration. As mentioned before, combining biomaterials with gene therapies could enhance tissue regeneration and modulate the immune response.

The imparting of biomaterials with specific architecture assures cell migration and blood supply is a crucial issue in tissue regeneration. Scaffold substrate plays a vital role in regenerative medicine, and its precise architecture is very important. Accordingly, engineers in cooperation with materials scientists are trying to develop traditional methods and discover new techniques for scaffold fabrication. Among these techniques, 3D bioprinting techniques are the most adequate technique for scaffold fabrication, specifically for cell-loaded and intelligent scaffolds. 3D bioprinting is a strongly suggested technique that can produce fully functional organs in the future. This was obvious from the increase in the number of publications concerning 3D bioprinting from 10 publications in 2010 to 1270 in 2023 based on the Scopus database. Ex vivo living tissue can be prepared directly by 3D bioprinting by including cells, growth factors, biomolecules, and hydrogel. Also, organoids, microenvironments, and organ-on-chip can be prepared by 3D bioprinting and have successfully used for drug discovery and cancer treatment can be fabricated by 3D bioprinting. The scientists hope to succeed in the utilization of this technique to fabricate scaffolds that mimic the structure of more complicated tissues, such as the liver, lungs, and narrow vessels (Fig. 5.8) with proper biochemical, biomechanical, and physiological properties. However, there have been already numerous successful semi-biological human tissues fabricated by 3D bioprinting technique, and there are parts of the cardiovascular system (e.g. blood vessels and heart) already present in the market for the treatment of myocardial infarction.

Some chronic illnesses need urgently organ transplantation as an obligatory choice for treatment, kidneys and liver are the most common organ transplantations. 3D bioprinting is hopefully a success to achieve organ-on-demand fabrication which can solve the problem of organ shortage. In the USA only about 122,000 patients were waiting for organ transplantation in 2016 [30]. Accordingly, illegal organ trafficking is activated in the dark market. About, illegal 12,000 organ transplants occur each year due to the scarcity of legal organ sources, and its trade reaches about 1.5 billion USD/year [31]. The most exploited people in organ trafficking are poor and vulnerable populations, such as migrants, unemployed, and homeless people. Besides, organ transplantation surgeries are expensive, and in some cases, they are defeating. The popularity of organ trafficking is considered obvious and evident in the field of tissue regeneration to replace organs totally (particularly for liver and kidney) is still in its infancy, and it needs more efforts and improvements.

Fabrication of artificial liver by 3D bioprinting remains a difficult assignment. That is because the liver is composed of different types of cells; liver sinusoidal endothelial cells, stellate cells, extracellular matrix, and hepatic cells. Hepatic cells form about 80–85% of liver cells, and their availability is limited. Moreover, hepatic cells lose

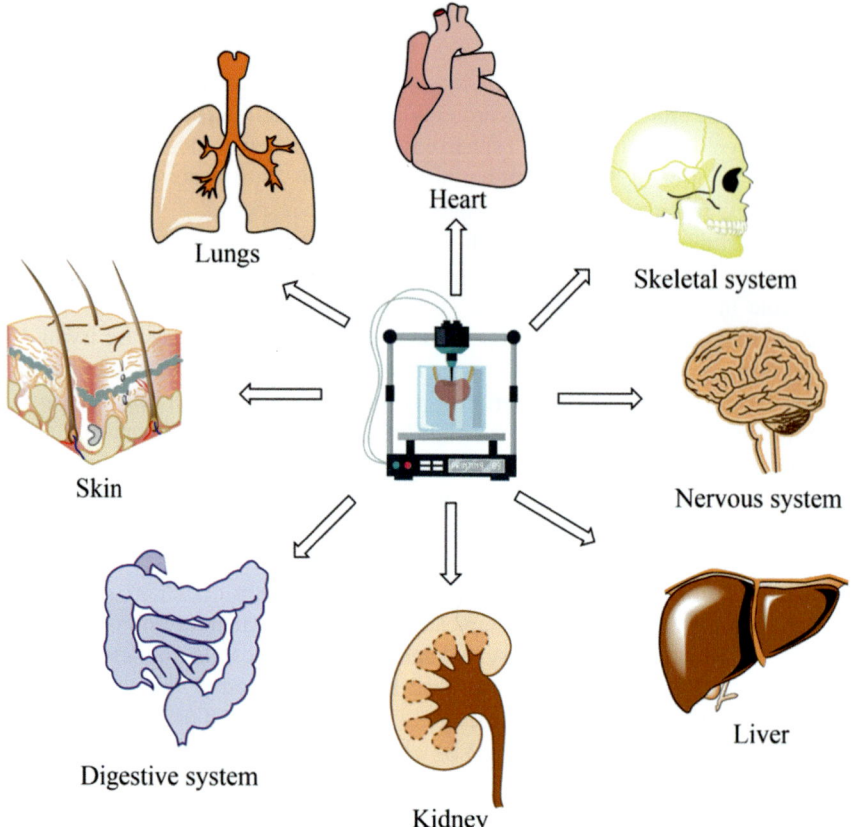

Fig. 5.8 Different organs that hopefully to be fabricated by 3D bioprinting

their function gradually when cultured ex vivo [32, 33]. Finally, the liver possesses a complicated system of hepatic blood flow. All of these reasons make it difficult to fabricate the unique architecture of the liver, combine different cell types, and keep their function during ex-vivo culturing. Likewise, the kidney possesses a complicated architecture and about 16 different highly specialized epithelial cells [34]. The kidney is composed of functional units called nephrons. Nephron is responsible for blood filtration including recovery of important nutrients, osmolarity regulation, and electrolyte balance. It has a complicated structure and combination of several types of cells which is considered a vast challenge to gain a fully functional kidney by 3D bioprinting [35].

Recently, innovative solutions proposed to achieve the desired goals in preparing fully functional organs by 3D bioprinting technique. Firstly, the printing resolution of 3D printing machine itself is improved to fabricate structures with microarrays that mimic the vascular structure of complex organs, such as the kidney, and the biomaterial precursors are modified to get organogenesis templates with desirable mechanical [36, 37]. Bioink is a main biomaterial precursor used to achieve a successful scaffold printing process. Different physicochemical properties can be controlled by the crosslinking step of bioinks. In addition, the application of smart technology on biomaterials by connecting them with internet remote control sensors to monitor, collect updated data, and assist the patient at a suitable time is believed to be an effective tool in tissue regeneration. Using of stem cells is considered an effective solution in tissue engineering. Stem cells are characterized by their regenerative nature, ability to self-renew, many times replication, differentiation into several cell types, and immunomodulation and secretion of useful cytokines [38]. This makes stem cells able to repair organ and tissue injured parts which is beneficial in regenerative medicine. Stem cells are categorized according to their source, such as bone marrow, adipose, amniotic fluid, umbilical cord, placenta, peripheral blood, and dental pulp-derived stem cells. However, the application of stem cells in tissue engineering is still limited due to their low differentiation efficiency, and differentiation into specific cell lineage is difficult. Therefore, growth factors have been applied effectively to stimulate stem cell differentiation. Moreover, stem cells can be modified by specific genes (by viral or nonviral methods, as mentioned before) that possess the ability to regenerate tissue. In addition, using of stem cell spheroids enlarges their potential role in tissue regeneration. That is because cell spheroids possess a high in vivo survival rate and closely mimic the host tissue. Lastly, AI has integrated increasingly into regenerative medicine. It saves preclinical experimental time due to its ability to predict material tissue response based on supplied big data, analysis of medical images diagnostic skill, capability to design precise geometry of 3D scaffolds and create successful models for drug discovery research. These innovative technologies are presented in Fig. 5.9 which will shape the next generation of tissue engineering, and they highlight the way to a promising future to fabricate fully functional organs.

Despite efforts made by tissue engineering scientists to fabricate fully functional organs or even parts of organs, they are still far from achieving this desired goal, which will prevent the suffering of many people from obtaining effective treatment for their diseases and eliminate organ trafficking. This leads us to know the fact that these tissues and organs were created by a powerful Creator who cannot be matched in His creation. Therefore, we will quickly discuss that the science of tissue engineering tells us that there is a Creator for this universe.

5.4 Tissue Engineering Tells Us There is a Creator

Away from religious, racial, or political tendencies, we will address here a topic that we touched upon while writing this book, and let us think in a loud voice. When we looked closely at the mechanism of the performance of every tissue and organ in the human body, we found that they are tightly built and perform their function with great precision. Not only that, but they communicate with each other to work

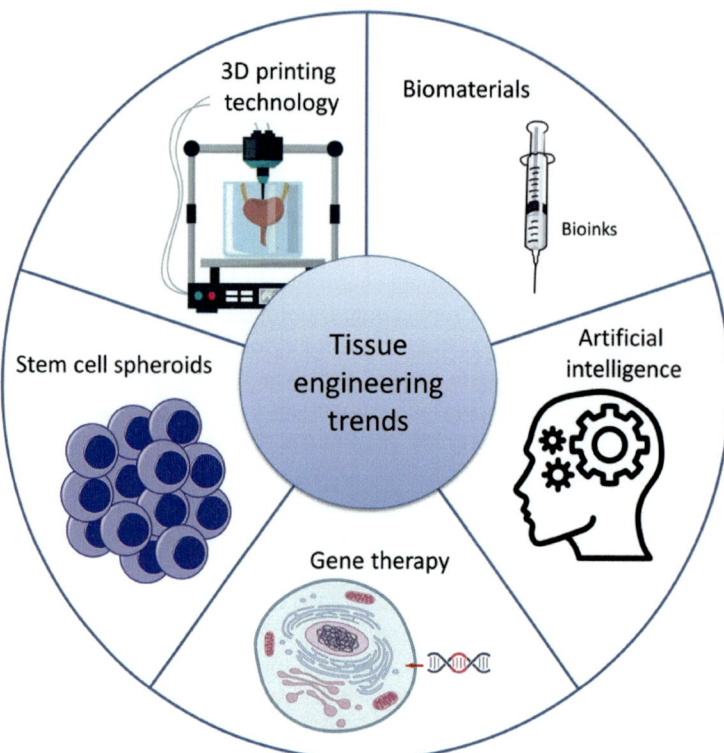

Fig. 5.9 New trends are used to improve the tissue regeneration approach

together in harmony and an accurate system without error, that is in case they are all healthy and not afflicted with any disease. Let us for example see how does cell work which is the building unit of our body. Cells differ in type, size, and shape according to their function in the body. Generally, they are composed of three parts; from outside to inside, the cell membrane, the cytoplasm, and the nucleus (Fig. 5.10). The cell membrane is a double layer of phospholipid molecules. It responsible for structural to maintain cell integrity, as well as, it controls the migration of materials from and to the cell via identification markers present in the cell wall. The nucleus determines the function of the cell and it is considered as a console unit in the cell. It contains 23 thread pairs of chromosomes which are composed of deoxyribonucleic acid (DNA). Part of DNA forms genes that define genetic traits. The nucleus also contains the nucleolus which is composed of ribonucleic acid (RNA). The cytoplasm contains a large number of very minute structures called organelles (e.g. ribosomes, mitochondrion, lysosomes endoplasmic reticulum, and Golgi apparatus). It takes charge of all cell functions, such as growth, expansion, replication, and synthesis of proteins. However, the functions of many organelles are still unknown.

The cell divides in a multiplication way either in the tissue rebuild process or in the embryonic formation stages. In the embryo, the cells differentiate into specific

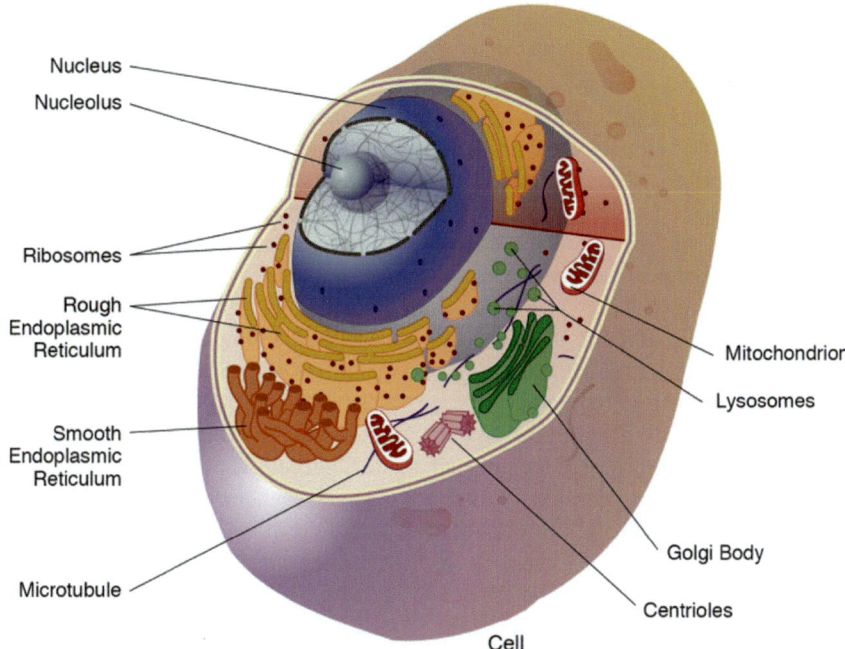

Nucleus

Nucleolus

Ribosomes

Rough
Endoplasmic
Reticulum

Smooth
Endoplasmic
Reticulum

Microtubule

Mitochondrion

Lysosomes

Golgi Body

Centrioles

Cell

Fig. 5.10 Schematic presentation of basic human cell structure [39]. From National Human Genome Research Institute which is free to use and redistribute

types during the division step and form functional organs. This process may involve a radical change in the shape of the cells during their differentiation, but the genetic material remains the same, with very limited exceptions. Cell differentiation includes a change in shape and some vital functions of the cell to suit the vital function that the cell must perform. There are many types of cells that a cell can differentiate into. If the cell can differentiate into several types, we call it a pluripotent cell. The most prominent example of these cells is stem cells. The other type of cell that can transform into any type of cell type is totipotent cells which are limited to the zygote and early embryonic cells.

Each specialized cell expresses a subset of genes. All of these genes constitute the genome properties. Cell types can be distinguished by their pattern of gene expression, and so, cell differentiation occurs by the transformation of gene expression from one pattern to another via gene and protein regulatory networks. An alternative mechanism of cell differentiation has been suggested by scientists who believe in Darwin's theory which proposed that cell differentiation occurs by selective and spontaneous gene expression. Hence, protein and gene networks are consequences of cellular processes rather than their causes. Darwin's has proven its failure in the light of modern science. Where, the science developed greatly and showed that the details of the life of living organisms are complex in their design, and it is not as claimed by the supporters of the theory of evolution, but it is the opposite. They claimed that the living cell structure is very simple, and the cell can be made by providing the necessary chemical materials for that, after a period it can be obtained. Based on this hypothesis, the Boeing 747 airplane can be made from iron scrap by a violent storm which combined and attached scrap parties and finally produced this airplane, this is impossible and unbelievable. Nevertheless, modern techniques, such as electron microscopes, showed that the living cell possesses a complex programming and structure such that it cannot be a coincidence to a degree that great scientific development did not, and will not, be able to fabricate a single living cell in any factory or laboratory. Accordingly, living cells cannot have come into existence by chance as the theory claimed. Likewise, the machine cannot be formed by itself; there must be someone to manufacture and produce this machine. Living cells, and hence all living beings, must have a creator and maker, and this is evidence of the existence of God who has the ability to create.

Once more, the tissue engineering field combines the living cells with a scaffold that mimics the architecture of the targeted organ. Stem cells are widely used in tissue regeneration due to their ability to differentiate and regenerate, but differentiation is not completely controlled to obtain zonal variation of cell density and type in the same tissue, such as the liver, kidney, blood vessels, etc. As well as, the precise structure, specifically the vascular system, inside the tissue and organ still needs novel techniques to achieve it. In this respect, the fabrication of a new fully functional organ can only be made by a Creator who absolutely knows the structure and composition of the living cell because he created it and can control its differentiation and gene expression. This does not mean that we should not research and develop the tissue engineering field, but we must know that man has limits in science and that the Creator can do anything.

References

1. Institute NC. https://www.cancer.gov/publications/dictionaries/genetics-dictionary/def/gene
2. Gonzalez-Fernandez T, Kelly DJ, O'Brien FJ (2018) Controlled non-viral gene delivery in cartilage and bone repair: current strategies and future directions. Adv Ther 1:1800038
3. Tang X, Sun C (2020) The roles of microRNAs in neural regenerative medicine. Exp Neurol 332:113394
4. Cucchiarini M (2016) Human gene therapy: novel approaches to improve the current gene delivery systems. Discov Med 21:495–506
5. Miyoshi H, Stappenbeck TS (2013) In vitro expansion and genetic modification of gastrointestinal stem cells in spheroid culture. Nat Protoc 8:2471–2482
6. Carter K, Lee HJ, Na K-S, Fernandes-Cunha GM, Blanco IJ, Djalilian A et al (2019) Characterizing the impact of 2D and 3D culture conditions on the therapeutic effects of human mesenchymal stem cell secretome on corneal wound healing in vitro and ex vivo. Acta Biomater 99:247–257
7. Costa EC, de Melo-Diogo D, Moreira AF, Carvalho MP, Correia IJ (2018) Spheroids formation on non-adhesive surfaces by liquid overlay technique: considerations and practical approaches. Biotechnol J 13:1700417
8. Vadivelu RK, Kamble H, Shiddiky MJA, Nguyen N-T (2017) Microfluidic technology for the generation of cell spheroids and their applications. Micromachines 8:94

9. Carballo-Pedrares N, López-Seijas J, Miranda-Balbuena D, Lamas I, Yáñez J, Rey-Rico A (2023) Gene-activated hyaluronic acid-based cryogels for cartilage tissue engineering. J Control Release 362:606–619

10. Li L, Huang Y, Qin J, Honiball JR, Wen D, Xie X et al (2022) Development of a borosilicate bioactive glass scaffold incorporating calcitonin gene-related peptide for tissue engineering. Biomater Adv 138:212949

11. Carvalho BG, Vit FF, Carvalho HF, Han SW, de la Torre LG (2022) Layer-by-layer biomimetic microgels for 3D cell culture and nonviral gene delivery. Biomacromolecules 23:1545–1556

12. Lee J, Lee S, Ahmad T, Madhurakkat Perikamana SK, Lee J, Kim EM et al (2020) Human adipose-derived stem cell spheroids incorporating platelet-derived growth factor (PDGF) and bio-minerals for vascularized bone tissue engineering. Biomaterials 255:120192

13. Zhou Y, Wang F, Tang J, Nussinov R, Cheng F (2020) Artificial intelligence in COVID-19 drug repurposing. Lancet Digit Health 2:e667–e676

14. Ng WL, Chan A, Ong YS, Chua CK (2020) Deep learning for fabrication and maturation of 3D bioprinted tissues and organs. Virtual Phys Prototyping 15:340–358

15. Krittanawong C, Johnson KW, Rosenson RS, Wang Z, Aydar M, Baber U et al (2019) Deep learning for cardiovascular medicine: a practical primer. Eur Heart J 40:2058–2073

16. Mackay BS, Marshall K, Grant-Jacob JA, Kanczler J, Eason RW, Oreffo RO et al (2021) The future of bone regeneration: integrating AI into tissue engineering. Biomed Phys Eng Express 7:052002

17. Golebiewska A, Hau A-C, Oudin A, Stieber D, Yabo YA, Baus V et al (2020) Patient-derived organoids and orthotopic xenografts of primary and recurrent gliomas represent relevant patient avatars for precision oncology. Acta Neuropathol 140:919–949

18. Wu Q, Liu J, Wang X, Feng L, Wu J, Zhu X et al (2020) Organ-on-a-chip: recent breakthroughs and future prospects. Biomed Eng Online 19:1–19

19. Marsano A, Conficconi C, Lemme M, Occhetta P, Gaudiello E, Votta E et al (2016) Beating heart on a chip: a novel microfluidic platform to generate functional 3D cardiac microtissues. Lab Chip 16:599–610

20. Kong J, Lee H, Kim D, Han SK, Ha D, Shin K et al (2020) Network-based machine learning in colorectal and bladder organoid models predicts anti-cancer drug efficacy in patients. Nat Commun 11:5485

21. Fuenteslópez CV, McKitrick A, Corvi J, Ginebra M-P, Hakimi O (2023) Biomaterials text mining: a hands on comparative study of methods on polydioxanone biocompatibility. New Biotechnol 77:161–175

22. Lu L, Zhang J, Guan K, Zhou J, Yuan F, Guan Y (2022) Artificial neural network for cytocompatibility and antibacterial enhancement induced by femtosecond laser micro/nano structures. J Nanobiotechnol 20:365

23. Batoni E, Bonatti AF, De Maria C, Dalgarno K, Naseem R, Dianzani U et al (2023) A computational model for the release of bioactive molecules by the hydrolytic degradation of a functionalized polyester-based scaffold. Pharmaceutics 15:815

24. Zhou C, Li Z, Lu K, Liu Y, Xuan L, Mao H et al (2024) Advances in human organs-on-chips and applications for drug screening and personalized medicine. Fundam Res

25. Bermejillo Barrera MD, Franco-Martínez F, Díaz LA (2021) Artificial intelligence aided design of tissue engineering scaffolds employing virtual tomography and 3D convolutional neural networks. Materials 14:5278

26. Chen D, Sarkar S, Candia J, Florczyk SJ, Bodhak S, Driscoll MK et al (2016) Machine learning based methodology to identify cell shape phenotypes associated with microenvironmental cues. Biomaterials 104:104–118

27. Tourlomousis F, Jia C, Karydis T, Mershin A, Wang H, Kalyon DM et al (2019) Machine learning metrology of cell confinement in melt electrowritten three-dimensional biomaterial substrates. Microsyst Nanoeng 5:15

28. Kalasin S, Sangnuang P, Surareungchai W (2022) Intelligent wearable sensors interconnected with advanced wound dressing bandages for contactless chronic skin monitoring: artificial intelligence for predicting tissue regeneration. Anal Chem 94:6842–6852

29. Yang J, Zhang YS, Yue K, Khademhosseini A (2017) Cell-laden hydrogels for osteochondral and cartilage tissue engineering. Acta Biomater 57:1–25
30. Wu C, Wang B, Zhang C, Wysk RA, Chen YW (2017) Bioprinting: an assessment based on manufacturing readiness levels. Crit Rev Biotechnol 37:333–354
31. Integrity GF (2017) http://www.gfintegrity.org/wp-content/uploads/2017/03/Transnational_C rime-final.pdf
32. Ma X, Qu X, Zhu W, Li YS, Yuan S, Zhang H et al (2016) Deterministically patterned biomimetic human iPSC-derived hepatic model via rapid 3D bioprinting. Proc Natl Acad Sci U S A 113:2206–2211
33. Kim Y, Kang K, Jeong J, Paik SS, Kim JS, Park SA et al (2017) Three-dimensional (3D) printing of mouse primary hepatocytes to generate 3D hepatic structure. Ann Surg Treat Res 92:67–72
34. Balzer MS, Rohacs T, Susztak K (2022) How many cell types are in the kidney and what do they do? Annu Rev Physiol 84:507–531
35. Reint G, Rak-Raszewska A, Vainio SJ (2017) Kidney development and perspectives for organ engineering. Cell Tissue Res 369:171–183
36. Levato R, Visser J, Planell JA, Engel E, Malda J, Mateos-Timoneda MA (2014) Biofabrication of tissue constructs by 3D bioprinting of cell-laden microcarriers. Biofabrication 6:035020
37. Daly AC, Cunniffe GM, Sathy BN, Jeon O, Alsberg E, Kelly DJ (2016) 3D Bioprinting of developmentally inspired templates for whole bone organ engineering. Adv Healthc Mater 5:2353–2362
38. Chen M, Przyborowski M, Berthiaume F (2009) Stem cells for skin tissue engineering and wound healing. Crit Rev Biomed Eng 37
39. Institute NHGR. https://www.genome.gov/genetics-glossary/Organelle

GPSR Compliance

*The European Union's (EU) General Product Safety Regulation (GPSR)
is a set of rules that requires consumer products to be safe and our
obligations to ensure this.*

*If you have any concerns about our products, you can contact us on
ProductSafety@springernature.com*

In case Publisher is established outside the EU, the EU authorized
representative is:

Springer Nature Customer Service Center GmbH
Europaplatz 3
69115 Heidelberg, Germany

Batch number: 08252336

Printed by Printforce, the Netherlands